Gulf Star
45

Also by Kim Bartlett

THE FINEST KIND

Gulf Star 45

KIM BARTLETT

W · W · NORTON & COMPANY · New York   London

Many personal and corporate names have been changed
to protect the privacy of individuals whom I met and liked
during my stint on what I have called *Gulf Star 45*.

Library of Congress Cataloging in Publication Data
Bartlett, Kim.
  Gulf Star 45.

  1.Oil well drilling, Submarine.   2.Petroleum
workers.   I.Title.
TN871.3.B37   1979   622'.33'8   79-16251
ISBN 0-393-01265-4

1 2 3 4 5 6 7 8 9

To Kax—who has always been there

Contents

Cast of Characters

Alton Otis McCall   *mechanic*
Baxter   *roughneck*
Bear   *roughneck*
Big Daddy   *roughneck*
Bullrider   *night watchstander*
Byron   *roustabout*
Carney Ewing   *barge engineer*
Catfish   *roughneck*
Clayton   *roustabout*
Clyde McMurray   *night driller*
Dan Walton   *night toolpusher*
Dave   *electrician*
Dwayne   *roustabout*
Elmo   *Company Man*
Freeman   *roustabout*
Galen   *mud engineer*
Grady   *day toolpusher;*
        a/k/a James Austin Stantley
Hank   *roughneck*
Harmon   *galleyhand*
Hippy   *roughneck*
Jackson   *steward/cook*

Jake Pitre *crane operator*
Jimmy *electrician*
John Nason *roughneck*
Jud *day watchstander*
Leroy Gaston *toolpusher*
Lew Millard *crane operator*
Nick *roustabout*
Pop *roustabout; a/k/a H.J. Carson*
Stokes, Charlie *day driller*
Tiny Long *crane operator*
Wade *crane operator*

Acknowledgments

You don't write books, especially ones like this, without the help of a lot of people. As you make your way deeper and deeper into the forest, becoming evermore overwhelmed by the trees, you never know who will appear to point you in the right direction. It is only when you are out the other side and can look back that you realize that not only have you made it through but along the way you've picked up a bunch of friends who, in the final analysis, are as much a reward as the product itself.

So it was with *Gulf Star 45*. Dirk Wiersma gave me the idea years ago. Ben Martin steered me away from the North Sea and down to the Gulf of Mexico. Corky Badgley became my "genii." Richard Hebert "gave me a chance." Bob Tannen did know the right people. Nancy Saucier found room for me at the inn. The Sons and Daughters of the Apocolypse—Lynn, David, Bill, Bill, and Walker—made room for me at their table. And my fellow "oil field trash" quickly forgot I was the "book writer" and let me be one of them.

At the end entered Laurie Matson and Doris and Tony Bocelle, who burned the midnight oil to make the manuscript presentable. And through it all there were Tudy, Dirk, and Jake, who had to put up without me for so long and then with me for that much longer while we got it all together.

There are others. They know who they are. To all of them, thanks.

Gulf Star
45

1

On the Road

It's raining. I can hear it beating on the sheet metal roof of the
balcony, splattering on the cobblestones of the small courtyard
two floors below. There's the low groan of a fog horn . . . another: freighters? barges? pushing through the night on the
Mississippi River as it doglegs around Algiers and straightens its
course through Plaquemines Parish to the Gulf of Mexico. And
despite the rain there is a burst of laughter from a group passing
down Chartres Street. An all-night New Orleans sound, it rises
over the roof of the Pontalba building where I live.

I throw my duffel bag over my shoulder, pull the door
locked, and descend the worn, broad stairs to the dank corridor,
past the layers of crumbling plaster and peeling paint, to the
old, heavy wood door, out to Jackson Square, almost abandoned
save for some single gays lounging on double benches and a
couple embracing in front of St. Louis Cathedral.

I turn left past the closed Café Pontalba, the chairs inside
turned upside down on the round tables, then left down Chartres, staring for a moment up St. Peter's at the ever-milling
crowd outside Pat O'Brien's Bar. I can faintly hear the nightly

duel of dixieland bands as they berate each other through open doors across Bourbon Street at the St. Peter's intersection. I step aside for a group of out-of-towners, New-Orleans-drunk and carefree, convention tags still pinned to their shirts.

Now left down Wilkinson Street with its set of empty store-fronts and double doors behind which the Jackson Square artists park their wagons, and finally to my little orange pickup—as ever, illegally parked. I toss my bag onto the front seat, take out my thermos of coffee and my can of peanuts, set them beside me, turn on the ignition, flip on the windshield wipers, and ease into the street; down to Decatur. I take a right at the Jax Beer Factory, straight up North Peters, past the low-life Greek bars to Canal. Right on Canal, slowly accelerating, timing the stop lights as they turn green, to Clairborne where I make an illegal left and follow it to the up-ramp for Interstate 10.

The Superdome is to my left. It is darkened. The last of the New Orleans Jazz-Portland Trailblazers crowd has vanished. I bear right at the Baton Rouge sign, pick out a handful of peanuts, turn on the radio, and settle back into the seat. It is 12:30 A.M. Behind me it is still Tuesday night; before me it is Wednesday morning. I am leaving the lights and fever of the city. Waiting for me is a dimly lit patch of rutted mud call Sabine Pass, Texas. Between the two is a five and a half hour drive alone through the rain. This time tomorrow I'll be one hundred and twenty-five miles out in the Gulf of Mexico, half way through the first of seven, twelve-hour work shifts, on a roaring iron island known as *Gulf Star 45.*

In the darkness I punch buttons until I home-in on "Charlie Douglas and the Road Gang," WWL out of New Orleans. For the next few hours I'm prepared to give a damn about the road conditions in Fargo, North Dakota, and to let my heart go out to Caroline from Tuscaloosa, Alabama, whose husband is hauling through a snowstorm into Dayton, Ohio; she wants Don Williams' "I've Got a Winner in You" played for him. Good choice. Nice Song.

I settle back into the seat, set my accelerator foot at sixty-five, and gradually submit to the arching beat of the windshield wipers. If everything goes right, I calculate, I should be able to grab half an hour's nap before we head out to the rig. I'm feeling weary already. . . .

2

The Bank

The small traveling alarm clock I've placed on the dashboard rings and I open my eyes. I've been awake for the past five minutes, waiting for it, not wanting to move until I have to, feigning sleep in hopes of fooling my road-weary body into a few more minutes of rest.

I don't have to open my eyes to know that the weather is still lousy, dismally rotten. Overcast, low-ceilinged, it is no longer raining outright, but threatening like a hairy fist held inches before the face. What can you expect? It has to be. Wednesday is crew change for *Gulf Star 45,* and ever since I have worked on the rig Wednesdays can be counted upon to be downright foul.

Depending upon which end of the change you are on, that foulness has its own particular import. One hundred and twenty-five miles out in the Gulf of Mexico there are forty weary men with seven days and eighty-four long working hours behind them. They don't want to spend an extra second out there. They want to "get to the bank" and on the road home. Bad weather can mean no helicopter; at worst, no crew boat. Oil rigs rarely stop drilling. No helicopter, no crew boat, no crew change, more work. The equation is simple, the thought is intolerable.

For us on the bank the situation is more subtle. Half of us will be working days this hitch; the other half, nights. The day shift, or tour (pronounced: "tower"), begins at 5:30 A.M. and ends at 5:30 P.M., when the night crew comes up. As far as the day crew is concerned, they have already begun work and the longer it takes to get out to the rig, the less work they have before them. For the night shift—and this includes me—the sooner we get out, the longer we'll have to sleep; the later, the less.

And that sleep is vital. Most of us have had six- to eight-hour drives; a few, ten to twelve hours. One roughneck comes from Atlanta, Georgia, fourteen hours away. Most of us have put in a full day at home before leaving for the rig. Most of us travel alone; those who traveled together may have slept some in the car. If we reached Sabine Pass early, we might have managed to catch an hour or so before mustering at 6:30 A.M. But if we have had any sleep it has been short and restless, curled up in a car seat—not the kind you want to rely on while working a twelve-hour tour on the drill floor.

I tug my duffel bag up from the floor, reach over and lock the passenger door, take the keys from the ignition, and push backward out of the truck, slamming the door locked.

Diagonally across the mud-skimmed road are two once white, now weather-beaten trailers. Both belong to the Sun Oil Company and function as gathering stations for the oil hands waiting for the helicopters to take them out to the various rigs. Sun Oil controls a number of blocks in the Gulf and is a silent partner in West Cameron Block 6— with the Meta Petroleum Company for whom we, of the Gulf Star Drilling Company, are working. The permanence of the Sun Oil Company in that part of the Gulf is reflected in the trailers which, though by definition are temporary and moveable, seem to have become eternally rooted in the East Texas mud.

I recognize Hippy's Chevelle planted across the road from the last trailer. On my way, I peek in. Hippy is sound asleep, a blanket pulled over him, but I can see he has on his light blue denim jump suit which tells me Hippy did some dancing at the Disco Club in Baton Rouge before he left. Hippy considers himself quite a figure in this suit; he announces proudly to anyone that it cost him seventy-five dollars. There are those who feel he should have saved his money.

I let him sleep. Someone will wake him when the chopper arrives. I cross the road, tightrope over the narrow bouncy board which spans the ditch, and climb the tacked-together wooden stairs that lead up to the door of the trailer.

The trailer is filled with listless bodies. They lounge at varying degrees ranging from the perpendicular along the plank bench which runs the back length of the trailer. Most are asleep, their chins on their chests, heads canted to one side, some resting on the shoulders of the next man over. Legs are stretched out straight, some parallel, some crossed; others are tucked under the bench. Mouths are slack and open, others tight-lipped. Bullrider is in the corner, his broad-brimmed, black bullriding hat pulled down over his eyes, the brim resting on his nose. He is gently snoring.

I nod to the few men who are awake but still fixed with what I've come to call the "oil field stare"; a vacant, distracted, day-dreaming look which takes in little of the surroundings but hints of fields and streams and coondogs yelping and fish jumping and the crack of a rifle—or, I sometimes suspect, nothing at all. The "stare" is particularly pronounced during coffee breaks on the rig when there can be as many as forty men all wide awake, each with a cup of coffee in his hands, each staring blankly to the front, all perfectly silent for minutes at a time.

Next to the door lies a pile of duffel bags and suitcases dropped on and around each other. I drop mine, it rolls down off the top. I continue to a wooden stand in the middle of the room on which a pad of paper lies. At the top is written: "First Flight," with a number of names already scribbled in. I count them, mostly for the exercise: thirteen, one more than will be allowed on the chopper. I flip the first page over. On the next —"Second Flight"—there are fewer names, so I write mine in along with my weight and the weight of my duffel bag. Again, this is an exercise. I should continue to the third page, but maybe this time—.

The lists are all part of a ritual which goes on at both ends of every crew change. It is happening right now on the rig where, owing to the crew's eagerness to get off, an element of viciousness is added. Each hitch includes roughly forty men, from galley hands to toolpushers. There is nothing lower than a galley hand and nothing higher than the toolpusher, who is the big boss. Between the two runs a hierarchy which in many cases is more subtle than obvious. For instance it is not clear who is more important: the driller who is one step in promotion behind the toolpusher, or the barge engineer whose job it is to keep the rig afloat and in balance so that the driller can do his work. Crane operators have more responsibility than rough-necks, which should place them somewhat above roughnecks and certainly above watchstanders, who assist the barge engi-neer. But where in this scheme do the mechanics, who maintain all the engines, and the electricians, who keep up the power, stand?—certainly higher than the roustabouts, who are a gigan-tic step above galley hands but substantially lower than every-body else.

Now, there are two ways the exchange of crews can be made: by crew boat or by helicopter. If the crew boat has to be used, the hierarchy question becomes moot, and everyone goes out together. Helicopters, however, have a limited capacity, and the number of crew they can take depends on the weather conditions, the size of the chopper, and the aggregate weight of the men chosen to ride. Under optimum conditions, and flying the large helicopter, twelve men can go at one time. Since for some reason only one chopper is usually used for crew change, and since the one-hundred-and-twenty-five-mile flight out takes about two and a half hours round trip, one's position on the scale assumes a major importance.

Taken objectively, the first in line to go out should be the day shift, who will relieve those who have just come off nights. But while this makes some sense and is in theory how it works, drillers and crane operators, regardless of their shift, assume the right to precede roughnecks and roustabouts. In the same way, roughnecks consider themselves to be more essential than roustabouts, while mechanics and electricians regard themselves on at least a par with roughnecks. Meanwhile, watchstanders, while performing a fairly ancillary position on the rig, entertain white-collar illusions and try to ride the barge engineer's coattails onto earlier flights.

I can view this ritual scramble with a certain dispassion because I am a roustabout and a "worm." What's more, I'm scheduled to work nights. The only time a roustabout could ever ride the first chopper is coming in, and then only if he is part of a car pool with the toolpusher, barge engineer, or driller. Day roustabouts will always precede the night shift and, therefore, generally take the second flight. Whenever possible, "worms" ride the third flight, for "worms," being the newest men and therefore the lowest next to galley hands, must give way to seniority.

But the fun has not really begun. Signing one of the lists is only to let it be known you are present and are one of the players. Soon now Grady Stantley, the day toolpusher, will push himself off the bench and mosey over to the stand and scan the names. Generally, he will fold the sheets back to an empty page and write: "First Flight," with his name at the top. He will then add the names he determines to be top priority until the first flight is filled. Then he will sit down. Or, if he is in a perverse mood, he will simply scratch out those names on the first list which he judges shouldn't be there and bring up those he wants. Whichever, that is the official list.

Now the jockeying begins. After a suitable wait, someone of middle to upper importance, usually Clyde McMurray or Charlie Stokes, the drillers, will casually make his way to the stand and will himself scratch and add names, making up the second list. This list is official only until the first flight has left the ground. Anything can happen once the bosses are gone. Knowing that they are on the first flight, Clyde and Stokes have merely tried to insure that their roughneck crews will make the second flight intact. Once the first flight has gone, it is not unusual for a crane operator to try to scratch a roughneck off and add one or two members of his roustabout crew, or a mechanic, feeling slighted, will bump off the last name and add his own, only to be bumped himself by an angry welder.

But at a certain point there are three lists. As with animals at play, there are instinctive rules which seem to defy common logic and ultimately prevail, and everyone returns to his seat resigned to his lot.

I find a place next to old Pop Carson. It has been a week since we last saw each other. During the upcoming week we will spend at least eighty-four hours together working and that many again sharing the same room. If this hitch is at all like the hitches past, we will spend half our time risking our lives together and depending on one another for our survival, the other half being bored ("fucking the dog," Pop calls it). I will keep him awake with my reading; he will scare my sleep away with his raucous snoring. We will share cigarettes, jokes, and bitches. Yet I don't shake his hand; he doesn't shake mine. We greet one another with his moving over and my taking that space, a simple welcome.

"How you doing, Pop?" I ask quietly, not wanting to wake Bear, who is dozing and nodding on my right.

"They ain't had a flight outa here in two days," Pop answers.

For Pop, that is an answer to how he is. He is part little old lady, part perfect servant. He worries about everything, personally assumes to a point of obsession any burden that's around, then obligingly submits to any and all decisions and orders handed down to him.

I can tell Pop is worried by the way he is smoking his cigarette. Quick, little puffs. No inhale. The butt held between his thumb and forefinger, like tweezers, the filter placed just inside the lips at the direct center of his mouth. The first time I saw him do this was coming in from the rig by helicopter into seventy-knot winds and a cloud ceiling so low the chopper was skimming the tops of the rigs along the way. The next time was right after Bear nearly killed both of us.

"Then we'll be going out by crew boat, it looks."

"They say the Gulf's running twelve-foot seas. They'd never send one of them little boats out in that."

"Hell, you say!" mumbles Jake Pitre, the crane operator, who is sitting on the other side of Pop. "Those crazy coon-asses will go out in anything."

"Well, there wasn't nothing outa here yesterday," Pop states with a dubious certainty. How does Pop know these things? It's only six-thirty in the morning! I'm aware he usually gets down to Sabine Pass early so he can get a couple of hours sleep before going out to the rig. But then he always sleeps in his car.

The truth is, Pop generally doesn't know what he's talking about—his is the stuff with which rumor mills are fueled. He turns whispers into shouts and somehow over the course of the past hour he has heard something about the weather and has successfully worried it into a set of hard facts.

"They'll fly us out if they can," Jake says. Part statement, part speculation, part wishful thinking. No one likes the crew boat. They'd rather risk their lives in a helicopter than endure the tossing sea. Two hitches ago I watched a man give up his job and go home when told he would have to go out to a rig by boat.

One thing is for certain, the rumor mill is back at work. For the next seven days it will be running full steam, encompassing about everything that has to do with the daily working of the rig. Everyone knows something; few know the same thing. There is a presumption that someone knows everything, but the better acquainted you become with the rumor mill, the less confidence you have that that one person exists. Instead you come to realize that if you wait long enough, whatever is going to happen will, and the hows and whys are incidental.

I look over at Grady, the toolpusher. If there is anyone in the crew who should know what is going on, it's Grady. He is sound asleep. He's been in the oil fields too long to be bothered. Either we will or we won't go out. If we don't go out today, we'll go tomorrow. The rig isn't going to stop drilling. Once I imagined he could find out what was happening if he wanted to. Then I learned that one of the powers he held over all of us was that of either not knowing or withholding when he did know. That is the boss's privilege, and Grady enjoys exercising it.

"You seen Freeman?" I ask Pop. Freeman is one of the four roustabouts in Pop's and my crew. Bear is the other. I know Bear is there, his head is bouncing off my shoulder.

"I think I saw him earlier. I guess he's out sleeping in the car," Pop says.

"Freeman won't be down for a couple of days," Jake says. "He's got to be in court back in Jackson."

"I guess he came down and drove back," Pop argues illogically. Jackson, Mississippi, is nine hours away. Pop is confused because Freeman is black and Pop is one of those who, having not been raised with blacks, tends to find one indistinguishable from another. There are four blacks on the rig: Jackson, the cook; two roughnecks; and Freeman. Not one even vaguely resembles another, except to Pop.

"So we're going to run shorthanded."

"More'n that," Pop says, "The fat boy is going roughneck this hitch."

"Whose place is he taking?" I know where Hippy is, there's Clayton over in the corner, I saw Catfish's truck in the lot. That leaves only the Old Man and Hank from Atlanta missing.

"That little guy. I guess he hurt his back last hitch or something."

"And the Old Man ain't showed up neither," Jake adds. "They don't figure he's going to."

"What's wrong with the Old Man?" I ask.

"Who the hell knows? He never called in. Probably too drunk to," Jake suggests.

"Don't he have a brother pushing tools?" Pop asks. "He don't have to worry."

There go the nightly Bou-Ré games. No Old Man to feed the pot. Bou-Ré is an old Cajun card game brought out years ago to the oil fields. As far as I know that is the only place it is now played. Aside from being very fast and susceptible to as many rules as there are players, it is well known for the size of its pots which will run as high as three or four hundred dollars a hand; this, from a quarter ante.

The Old Man was an inveterate Bou-Ré player, the most regular of the regulars. He was known to play throughout the night, wash, eat, put in his twelve-hour tour, and return to the table with his dessert in his hand. I can't recall ever seeing him win a hand but I also never saw him leave a game. I once asked him how much money he brought out to the rig with him:

"I never count it. I figure I'd use the money to drink and fuck. I'm getting too old now so it's all right someone else drinks and fucks on my money. I just hope I got enough for a case of beer to get me home."

No one ever knew how old the Old Man was. Most guessed he was around forty-five, but everyone agreed he looked older than he was. They said he was a "damned good roustabout, too slow to be a roughneck, but a helluva good guy to work with." That's what they said when he was on the rig. They never mentioned his name again after he left.

As for "fat boy," Bear, going roughneck, it is to be expected. The only surprising thing is that it happened so fast. He only came on last hitch and then as a roustabout. To this day Bear will insist he doesn't know how he became a roughneck so quickly, but the truth is, it was inevitable. He was brought on by the night toolpusher, a family friend, and he comes from Crystal Springs, Mississippi, which is just up the road from New Hebron where Clyde, the driller, lives. He can easily transport Clyde to and from Sabine Pass.

I can tell by the disgust in Pop's voice that he is relieved by this development and I know Clyde, the driller, well enough to suspect he is elated.

Pop is a working man and has been for most of his fifty-three years, the bulk of them spent as a logger in the state of Washington. He's seen a lot of men come and go, and he has developed that detachment that comes from working side by side with an array of partners. Names and faces meld together; only places and incidents retain clarity—as stories. Pop has only one standard: a man works, otherwise he should be run off. Pop has his lazy ways, but he rarely stands around. Bear stands around. Pop doesn't like that. One day Bear made a mistake on the air hoist and sent a stack of thirty-foot drill pipe end over end off the catwalk down toward Pop and me. Each pipe weighs around five hundred pounds. They landed in some equipment ten feet from us and stuck there. If they had tipped over they would have killed us; if they had bounced, they'd have mauled us. Pop's rage was matched only by his fear. For the next fifteen minutes he couldn't tie a simple clove hitch; all he could do was mumble, "I guess it wasn't my day to die." He never forgave Bear.

As for Clyde, in one stroke he was rid of two thorns and had acquired what for him was a rose. Both the Old Man and Hank had been to college, although neither of them flaunted it. Both were hard workers and did what they were told. But the Old Man was older than Clyde and, worse, was fiercely independent. Hank was from Atlanta, Georgia, which in Clyde's mind made him a Yankee. It makes little difference that Bear admits to being "probably the dumbest guy on the rig." He's from Mississippi. That makes him a good ol' boy.

The men are straggling in. Some I know, some I don't. There are other crews going out to production rigs—specialists or mechanics with a quick job which may get them home by night. These men are wide awake. They talk loudly to one another. I watch members of the *Gulf Star* crew open sleep-dumb eyes, then twist around and try to fall back to sleep.

The pile of duffel bags and suitcases grows; a new pile is started. A young red-headed kid comes in. I've never seen him before. He looks at the list. "For *Gulf Star?*" he asks. "Yeh," says Carney Ewing, the barge engineer. The kid writes his name on the first list. A new hand, from the looks of him, a galley hand. He backs over to the second pile, slumps down, and takes a book from his blue jeans. I can barely see the title, but I know from the cover it is a Louis L'Amour. What else?

Suddenly the door is thrown open so hard it bangs against the outside of the trailer. The open space is filled with a huge shape.

"Good morning, Motherfuckers."

It's Big Daddy.

"How ya at, Big Daddy?" asks Carney.

"Shee-it!" the big form answers.

"You're already on the list, Big Daddy," Clyde contributes.

"That's good," he says. "When's the first flight comin'?"

Grady shrugs. "Maybe it ain't," Jake offers.

"Hasn't been one for two days," Pop contributes.

"Crew boat?" asks Big Daddy.

"Might have to be," says Stokes, Big Daddy's driller, from his seat in the corner.

"Motherfucker," Big Daddy sums up, and heaves himself over to the long bench where space has been made through compacting the rest of us so that the leaners like Bear have been forced upright—and awakened.

Big Daddy is a man with the command of and the need for very few words. His sheer bulk expresses him best. Before I ever met Big Daddy someone described him: "unbelievable shoulders . . . and wait 'til you see his ass." And wait 'til you see his arms and his thighs! Once out of curiosity and intimidation I checked his weight on the flight list—two hundred and fifty pounds, by his estimate; and he is probably no more than six feet tall.

Sheer testimony to his strength is his having made an appearance today. Last hitch he had his forearm crushed between two strings of drill pipe while his roughneck crew was coming out of the hole. Each string is three joints, or ninety feet, long and weighs at least fifteen hundred pounds. Big Daddy was out of commission for the rest of the hitch, but, as he is explaining right now, all that happened was "it swole up good." There was general agreement that the accident would have shattered anyone else's arm.

Big Daddy's entrance has temporarily "shut down" the production-rig hands. They are staring at him as he settles into his routine. Between his feet he has placed one of the empty coffee cans which lie around the trailer as combination ashtrays and spittoons. Out of his shirt pocket he takes a small, round can of Copenhagen snuff from which he daintily picks between thumb and forefinger a large snatch of snuff. This he dexterously locates between the lip and teeth of his prognathous jaw. The snuff in place, he lowers an almost bovine tongue as a lid. In a matter of moments the saliva has begun to work and Big Daddy leans forward, dropping his head toward his knees, and hurls a spit directly into the can. Only then does he raise his head and slowly take in the crowd of men in the trailer, fixing each with a benign glare. Quietly, what few conversations had been in progress before his explosive arrival are resumed, while Big Daddy retires into rumination.

It's now eight o'clock. If the helicopters are going to fly, we should start hearing them: the sound of gigantic wings beating as the propellors chop through the air, at first just a distant, quick thumping growing steadily until it seems to shake the whole trailer. The men are geared to dash to the door. The piles of bags will be ripped apart as those on the first flight haul what's theirs out from underneath. They will squeeze through the narrow doorway and down the stairs, through the mud around behind the trailer, past the bogged down cars in the lot, leaping the ditch to the field where the chopper will be settling down. Even before it comes to rest, the men, heads ducked, walking fast and low, will have opened the luggage compartment in the tail, shoved their gear in, and grabbed a seat, careful to leave the co-pilot's seat in front for the toolpusher. Life jackets will be passed around and struggled into, seat belts locked around waists. Then they will sit like good little boys on the school bus while the pilot checks over the flight list before lifting off. And those left behind will lounge against the side of the trailer, watching, then amble back to adjust, if possible, the next flight list, and return to sleep.

But eight o'clock passes. And eight-fifteen . . . eight-thirty. Once there was a beating sound and inside the trailer the men moved, but the sound disappeared and the men settled back. A feeling of gloom set in. With each succeeding minute, the crew boat possibility comes closer. At one point Catfish came in to report that it looked to him as though the ceiling was lifting. Clyde had gone out and looked and had reported, "Catfish is full of shit."

At around eight forty-five, rain can be heard beating lightly on the trailer roof and a few minutes later a young black man in a yellow foul-weather jacket, hood up, enters. He doesn't have to say anything. Everybody knows who he is, although few know his name. He's a "gopher" for Tom, the dispatcher in the Meta Petroleum trailer. There can be only one reason for his arrival.

"Get your stuff down to the *Sea Wolf*," he says.

"Motherfucker," says Big Daddy.

3

The Crew Boat

It is a desolate band of men slugging their way across the Air
Petroleum landing field, past the grounded helicopters, picking
their way around puddles, and making for the rutted, muddy
road leading down to the piers along the Sabine River.

There is little pleasing to the eye. When you say you leave
for the rigs from Sabine Pass, you don't mean from the four
corners back yonder a mile—that group of low slung café-bars,
roadside diners, and seemingly abandoned gas stations—you
mean this: the oil dump along the river next to the Coast Guard
station.

In the best of the oil field tradition it is a dump, a place
where stacks of old and new pipe lie rusting together, where
busted boards and discarded machinery are strewn about
in once conveniently vacant spots, where dry drilling mud,
cement, and chemicals from torn sacks and tanks have blown
about, settled, and mixed in the rain with the East Texas clay
to form a clinging slop. The constant traffic of cars, trucks,
cranes, and fork lifts has splattered it on the battered trailers
which function as the field homes for multi-million dollar corpo-

rations lodged luxuriously in elegant buildings like the Dresser and Jefferson Towers in Houston.

Once, before the oil companies came, the place must have been a docking area for barges to on- and off-load grain. There is a large, decrepit warehouse off to the left. Years before, Sabine Pass was the scene of a dramatic battle between the Mexicans and the soldiers of the Republic of Texas. There is a sign at the bend of Route 87 memorializing that moment in history when, as seems to have been the case in Texas, a handful took on the Mexicans and fought them to a standstill, thus saving Sabine Pass. For this?

The crew boats are tied to the left just beyond the Meta trailer. There are three of them—*Sea Dog, Sea Wolf,* and *Scorpion*—lined up like horses in the gate, sterns against the low dock, bows pointed into the river, black smoke puffing out the two stacks amidships, port and starboard. They are each painted white on dark green and belong to Black Gold Marine of New Orleans.

Crew boats have only one major function—to get a crew or a few pieces of equipment out to a rig as rapidly as possible. They are built a bit like PT-boats, long and slim, the wheelhouse and the cabin forward with a long afterdeck covered with heavy cypress planking. They are fifty to sixty feet long at the most and can cruise easily at twenty knots, which is what makes them very attractive to a certain group of younger southern Louisiana wildmen who Jake, the crane operator, earlier described as "crazy coon-asses." There is an old joke. It goes like this: "You know what a coon-ass compass is? A flashlight. He'll stick it on the bow and follow it wherever it goes." Stories abound about the captains of the crew boats, of how they race from rig to rig to find out where in the Gulf they are. There is only one thing I know for certain, the captains I've been with know only three gears—full forward, full astern, and full stop. Anything else is a mark of cowardice.

So it is with marked reluctance that we step off the dock,

over the stern bumper of rubber tires, and onto the afterdeck of the *Sea Wolf.* No sooner are we on board than a change comes over the *Gulf Star* crew. They become very quiet. Even Big Daddy spits out his chaw. They move with the wariness and resignation of sheep into the chutes. These are landsmen, and most of them from Mississippi. They are not used to the sea, to boats. They don't trust it, they don't like it. Helicopters are different: they are machines, and the air is a neutral space, neither land nor sea. You don't get killed in the air, generally; it's the land or the sea that does the job.

There are a few who are used to it, the older hands like Grady, Carney Ewing, and old Pop, and those like Dave, the electrician, and Hippy, who served in the navy. You can spot them. They have grabbed seats on the aisle, that is, over the keel where the rolling will be less pronounced.

Then there are others like Bear and Bullrider, Jud, the watchstander, and two kid roustabouts from Sebastapol, Mississippi—Dwayne and Byron—and Big Daddy, who have discovered early that the hatch to the crew's quarters below is open and there is a color television below.

The *Sea Wolf* is a relatively new crew boat, a little over a year old, and it is luxuriously fitted out. The upper cabin somewhat resembles a bus with an aisle down the middle between two rows of benches. The benches are moderately soft but slippery, covered with heavy orange vinyl. A man can curl up on one of them comfortably. Forward there is a set of short stairs leading through the bulkhead to the wheelhouse. To the rear of the cabin is a small bathroom.

Below the cabin is the boat crew's quarters. If the two large diesel engines and the generator were not roaring just behind the quarters, and if there were not a pervasive odor of diesel fuel, you could imagine yourself in an intimate night club, such is the decor. Low, soft benches down the sides. Triangular tables spaced along them with soft light emanating through white cylindrical shades. And a large color television placed in the corner. There is an aura of the good life about the lower deck of the *Sea Wolf,* Black Gold Marine's gesture to the four-man crew which mans the boat two weeks at a hitch.

It is unusual that the hatch to the crew's quarters is open. The boat crew is generally protective of its privacy, especially before an onslaught of oil field hands. There is little love lost between the two crews. We don't want to be there, and they don't want us. We are an inconvenience at best, and at worst a pain in their butt. They know as well as we do that the only reason we are on the boat is that the weather is too bad for the helicopters to fly. And we both know that while they are used to the heaving and wracking of the seas, we are not, and forty men cooped up under those conditions can lead to a mess which they, the boat crew, will ultimately have to clean up.

But neither crew has a choice in the matter. The Meta Petroleum Company pays Black Gold Marine $1,500 a day to have its boats available upon command. Meta is willing to pay because it is also paying the Gulf Star Drilling Company $22,000 a day to keep the rig drilling ahead. And Gulf Star is now paying us, although we aren't doing anything, while it is paying another forty men who are on the rig and doing what we should be doing. The longer we wait at Sabine Pass, the more it is costing Gulf Star, who in turn will pass the loss onto Meta which is already paying around $44,000 a day to explore Block 6—.

One hundred and twenty-five miles is a long boat ride. The question is, how long will it take to get out to the rig? Speculation ranges from six to eight hours. Which all depends, of course, on how rough the Gulf is. That is the unspoken question, and the answer comes quickly.

Steve, the lanky deck hand, appears at the head of the stairs with a long box of Glad bags. The sight of the box is enough. A groan rises from the oil hands, loud enough to bring Jud, the watchstander, up from below to see what the trouble is.

"I don't know if any of y'all get sick," Steve starts, "but it's a little rough we understand, and maybe y'all could use one of these."

"How rough?" yells someone in the back. From the squeak in the voice it sounds like Otis, the mechanic. These are the first words I've heard from Otis, which is amazing because Otis is known to carry on full conversations in his sleep.

"Weather report says eight- to ten-foot," Steve answers, ripping off the bags from the roll. "But they say the wind's goin' to blow up, so we'd sure appreciate y'all take one."

There is very little exercise of vanity. Just about everyone does take one, some take two. It's all right to be sick. Where you will get in trouble is if you do it on someone. The only person who passes a bag by is a new galley hand seated by the window across from me. He is an older man, appears to be about fifty although with galley hands it is often difficult to judge. Many of them appear to have spent a long life in a short time. Fortunately, it turns out, another new galley hand sitting next to him takes two. Fortunately for me, there is a narrow table between us.

The rear door to the cabin opens and a long, lean apparition with shoulder-length hair and a Bee-Gees T-shirt peeking out from a long, brown leather jacket, enters, clanging the iron door after him. He streams up the aisle toward the wheelhouse. I know him slightly. He's the captain. He sees Grady and stops to shake hands.

"How y'at, pardner?" he asks. He is one of the "Y'ats" from New Orleans, that younger set whose claim to status is the nasal twang to their "y'ats."

Grady tells him he's all right now but he's going to be better when we all get to the rig. He knows what Grady's talking about.

"Dig it," he says, laughing, and climbs into the wheelhouse. I know what is coming next—the tape into the stereo and the crashing, amped up music. Mercifully he shuts the wheelhouse door.

The twin diesels gun, the boat shakes. Both the "Today Show" from below and the rock music from the wheelhouse are drowned out. Slowly we pull out into the Sabine River, ease to starboard, and begin cruising toward the breakwater and the Gulf.

Through the fogged windows I glimpse the large supply boats—*Dearborn 206, Confidence, States Victory*—loading drill water and dry mud at the pier. The booms of the cranes stick up into the gray overcast. There's the Coast Guard station to starboard and a dredge working to port. A barge passing by sets up a roll for us. We pick up speed but just for a moment, then slow to a crawl as we pass by a fleet of shrimp boats moored by the mouth of the river. Then we are at three-quarter speed as we pass through the narrow channel by the rip-rapped break-water. And there it is, the roar and surge of full speed ahead and the heave and pitch of the first wave.

There is a plaintive "Oh, shit!" from the corner, a quiet groaning on both sides of the aisle, an involuntary adjusting of positions, and a settling in for what everyone now knows will be a torturous eight hours.

I know the *Sea Wolf* and what she will do in rough seas. Selfishly foresighted, I have commandeered an entire bench which I guard jealously. Being able to lie down and sleep is a whole lot better than having to sit up and knock heads and shoulders with the person next to you.

I can hear the haranguing rock music through the wheel-house bulkhead, but the television below has been off for hours, replaced now by sounds of a different order—a gagging and a moaning. The heat and stuffiness must be oppressive down there. Once in a while an ashen face peeps over the hatch. There is a gulping of our fresher air, and the face disappears. The two kid roustabouts have come up and snuggled into a place on the floor by the pile of bags in the far corner.

There's a steady parade of men to the bathroom. The visits are always just too long and the next in line starts weakly beat-ing at the door. "C'mon, man, hurry it up in there."

Suddenly across the table next to me I hear,

"Oh, my God . . . oh, my God. Where's a bag?"

It's the elderly galley hand, the one who didn't take a bag, the one who for the first hour had with such bravado listed all the ships he'd been on, the seas he'd sailed, and the ports he'd visited.

Wade, my crane operator, is sitting next to him. He has quickly opened his eyes and I can see he is making a rapid decision. Wade has a bag which he hasn't used yet, but . . . he is also on the aisle and can bail out at a moment's notice. What Wade will do depends on which way the galley hand turns his head, which is now faced towards the table.

The galley hand by the window does have two bags, but one is nearly half full, and the trip has at least three more hours to go. Since none of us has eaten since the previous evening, the chances of his having enough left in him to warrant a second bag are slim. Generously he passes the second bag over. Just in time.

The wretching sound is voluminous, the agony wrenching. The older galley hand's face is white, his neck red, the veins popping out. The effect is rippling. The young galley hand loses his control and with a skill born of hours of constant practice, buries his face in his bag. Wade, who has been paying close attention, feels an advancing queasiness and frantically digs his bag out from between his legs. "Jesus Christ!" groans Carney, the welder, from across the aisle. I feel none too sure of myself, suddenly, and duck down below the edge of the table and close my eyes.

The twisting, the banging, the roar, the pervasive diesel odor, the gagging, the mumbled cursing, they are all part of a rhythm which we in our state of suspended animation have come to endure. We are numb, lifeless, so many bodies in a pitching can. We are almost totally isolated, caught between the two worlds we know—the rig and the land. It doesn't help to wipe the steam from the windows. Waves washing over the top blur the sea which itself is gray under a heavy, low, gray mist. What time is it? How long have we been under way?

When will we get there? Where are we? Who knows? Who cares? Even the rumor mill has been stilled for the time being.

Then the engines cut back and the boat settles into a regular roll. This has happened before. I wait for the boat to ride up and up and down and the engines to gun forward again.

"There's the rig," someone yells.

I become aware of the crackling of the boat's radio, and I can hear the captain saying something. Mercifully, the music is off. The men are moving around in their seats, some standing up in the aisle. Out of the hatch looms Big Daddy mumbling something about getting "the fuck outa here." And here comes Bear after him, looking wan and, if possible, lighter.

"Twenty-three fucking times," he repeats to everyone around him. "Twenty-three times."

"Was that you made that mess down there, Bear?" asks Jud, the watchstander.

"You got that right, hoss," says Bear. "I'm here to tell you this was one sick Bear."

The mere mention of the rig has injected life into the crew. It is almost as though the crew boat trip hadn't happened. The way they are jostling to get in line and move out the rear cabin door is reminiscent of a high school football team arriving for an away game.

But the rig is curiously nowhere in sight. Out the windows on both sides stretch miles of uninterrupted rolling gray. Then, rising up on a wave in the distance is a white buoy, and off to starboard, appearing vaguely in the mist, is the outline of the Sun Oil production rig about a mile away. And the waves are no longer coming at the bow but are rolling into the stern. The engines race again but the boat moves in reverse, stuttering into the sea; through the now open door I can see one leg of the rig. From the way the boat is sitting, that leg practically fills the door, it is that enormous.

And down from the sky drops the basket. It hangs there, like a net, red life jackets stuck in the mesh, waving in the wind, while the crew boat maneuvers under it. With a straddling kind of walk, Grady moves towards it, grabs ahold of the mesh, steadies its swing, and, looking way up, makes a circular motion with his free hand. The basket drops to the deck, the mesh going slack. Immediately, Grady tugs out a life jacket and throws it to Carney Ewing. Clyde has already reached the basket, Stokes right behind him. They don't bother to take the snarl out of the jacket's straps. They slip their arms through the easiest possible way, forget the two buckles, and step onto the rim of the basket, grabbing hold of the mesh or sticking their arms through and locking their wrists. Grady raises an arm over his head, swings it around a couple of times, the slack goes out of the lines, and the basket jerks off the deck and straight out of sight. And the line of men advances four places.

For me nothing so completely captures the dangers implicit in working oil rigs as the basket. Even as it lies slack and innocuous on the rig deck, it reminds me of a sleeping rattlesnake, and now, as it dangles aloft, seventy feet up, swinging back and forth in the gusting wind, puny against the towering gray immensity of the rig itself, the red flaps of the life jackets are like snapping fangs. Have I really got to get on that thing? Is that really the only way to get from here up to there? I've asked the question a hundred times, in silence and aloud, always with the same answer: "That's it, pardner. Just get on and hold on, and whatever happens, don't let go."

"Don't let go." The advice has been deeply etched in my mind ever since Wade, the crane operator, told of the day he was lifting some roughnecks off a supply boat:

". . . and there was this new roustabout who didn't know his ass from a rat hole. And he's holding the basket steady 'cause the boat's moving around a lot and the Goddamned idiot is standing with his back to the stern, and you know how those boats don't have no stern.

"Well, he gives me the sign to take it up, and I give it a quick jerk to get it off because I know we got twelve-foot seas running . . . when just then the boat comes back up and takes off at the same time, and the basket hits this guy right about the chest and he grabs the bottom and he's hanging there. There's no boat, I got these seas so's I can't put the basket down in the water to let him climb on. I got to get him up cause if one of them waves hits him, he's into the water and no one's ever going to find him, he gone. . . . And I'm on one of those big jack-ups—with the one-hundred-foot lift—and I just decided right then: 'I'm going to bring the sonuvabitch up,' and here comes the basket with this guy dangling off the edge. We made it up, but you never seen a scared motherfucker like that in your life . . . and I'll tell you, brother, I got outa that crane and had me a good cup-a-coffee. . . ."

Suddenly there is no more time for worrying. The twin diesels are straining, pouring out heavy black smoke over us as the captain urges the crew boat back against the twelve-foot seas until the stern is beside the middle leg of the rig. The four two-thousand-horsepower generators on the rig are going full blast and their roar is magnified by the tunnel formed by the underbelly of the rig. The rain is slashing in, and waves, built higher by the backwater, are breaking over the stern and engulfing those of us standing on the deck waiting.

There's the basket suspended about ten feet just off the stern. Hippy is with me; we've moved to the middle of the deck, are leaning forward, legs wide apart; Pop is holding my duffel bag. Hippy makes his circling sign to the crane operator, his fist clenched, index finger pointing down. The basket drops lower, the captain races the diesels, the boat slips under the basket. Hippy and I run aft, each grabbing a shroud, and haul it as far forward as we can. Again the down signal and the basket collapses in a heap on the deck.

We yank at the life vests, toss one back to Pop, one to Otis, the mechanic, keep one apiece, and struggle into them, trying to keep our balance.

Pop and Otis come up; Pop hands me my duffel bag and makes an "up" sign to the crane operator. The basket makes a flabby cone. We shove our gear between the shrouds into the center, grab our spots across from one another, place one foot on the ring, holding the other on the deck for balance.

Pop is frantically signaling "up." So is Otis. Hippy is yelling, "Get the fucker outa here," as if the crane operator could hear him. I hardly can.

The cable jerks; we leap up, each setting his other foot; and the basket swings up and out and against the port stack. We hold on as we twist precariously around the stack and over the water and swing clear and out like a pendulum run amuck. And up the basket climbs, still swinging.

I take a last look at the sea and the boat below. I look at Hippy across from me.

"None of your games now, brother," I say to him. He smiles his macho leer at me. I know Hippy and personnel baskets, and he knows how I feel about them. I look at old Pop. He has a cigarette hanging out the corner of his mouth. Otis is standing ramrod straight, looking up at the approaching lifeboat, and already he's talking—about food.

4

First Tour

I

There are about forty men milling around the basket, all showered, shaved, and dressed for the "bank." They could be waiting for the last bus to town, the way they press in around us. Pop and I grab our gear and head for the far stairwell, Hippy and Otis to the near stairs. As we push through the men we say things like "How you doing?" and "Enjoy the trip back," and they say to us "Good, brother," and "God dah-yum."

There is only one person I know from this other crew, and I don't see Baxter's face anywhere. We met by chance my first hitch on the rig. He had laid over to help out a short roustabout crew. I've scarcely seen him since, yet we now share an apartment back in New Orleans. He leaves his part of the rent in an envelope on the desk.

Sometime before I came on the rig there were two brothers on it, both electricians. They worked opposite hitches, slept in the same bunks, lived in the same town, and saw each other only at crew change. I don't know the man I'm replacing. He may have been the one who answered "God dah-yum" to my "Enjoy the trip."

Pop and I shove out the other side, our backs to that crew. They are jockeying for the basket ride down, their backs to us. In fifteen minutes they will be gone. Already the rig is ours. We never bothered to ask how the week went, how much hole was made, did anyone get hurt. In a way it's none of our business —that was their week; we're beginning ours. We are two utterly separate crews, and we have completely different ways. We just happen to do the same thing on the same rig for the same company.

Pop and I are working nights this hitch which means we had better hurry down to our room and unpack if we want to get supper and some rest before starting. It is already past four o'clock, and the first tour begins at five-thirty. I can feel it in my legs—it's going to be a long night. Even old Pop looks a little bowed. Within the last day he has driven alone for ten hours, caught a couple hours sleep in the front seat of his car, taken an eight-hour crew-boat ride, and now, like the rest of us, has a twelve-hour hitch to put in. He says he slept some yesterday afternoon, but that was after having driven two hours to Pensacola, Florida, to visit his sister-in-law who is in the hospital with a heart attack, and two hours back.

"Stand up straight, Pop," I kid him as we head down the iron-grate stairs into the mud pump room.

"Just keep walking," he answers.

Working nights means our bunks are over on the starboard side with the galley hands'. Given the hierarchy of the crew, this is tantamount to segregation. Just about everybody else is quartered on the port side, including the night roughnecks. There is actually very little difference between the port and the starboard quarters, except that there are no toilet and shower facilities in the rooms to starboard, so we have to walk almost the length of the corridor to wash up. And we do have to cross to the port galley to eat. However, it is somewhat quieter in the starboard corridor which makes sleeping easier during the day. That can be an important difference.

The reason there are port and starboard quarters is that

Gulf Star 45 is designed for overseas work, and, by international
law, there must be separate facilities for the American and the
national crews on board. The law apparently does not stipulate
that they be equal, for besides the lack of private toilets and
showers—which the Americans get—the galley is smaller and
the waste treatment plant is located to starboard. No sooner do
you enter the corridor than you know it is there and how far
away you are from it.

"Geezuz!" I grumble as we enter the starboard corridor.

"What's the matter?" asks Pop.

"Smells like a shit house," I say. And it does. "Must be
having trouble with the system again."

"I can't smell nothin'," says Pop. In this instance he is lucky,
but the reason he can't smell anything at any time—and he's
curiously proud of the explanation—is that he "burned my
smeller out with whiskey years back." His lack of a "smeller"
saves him not only from the waste treatment plant but from the
diesel and oil smells that pervade almost every part of the rig.
It also deprives him of the taste of food. Pop is one of the few
people on board with no complaints about the meals that Jackson, the steward, cooks.

"You going to eat first?" I ask. I'm starving. Anything Jackson has for us tonight is going to be good.

"I'll stow my gear first," Pop says and turns right, through
the bathroom to the locker room aft to get his work clothes. Pop
is businesslike. He doesn't feel at ease on the rig until he has his
work gear settled around him and his work clothes on.

I continue forward to the cabin marked "night roustabouts." The door is open. I have my choice of bunks. I throw
my duffel bag on the lower right, reach in, and flick on the night
light over the head of the bed. It's all force of habit. If Pop had
been first, he would have taken the lower left. If Freeman had
been with us, he'd have taken the upper left. We've worked this
system out. The bunk over mine is always left empty, so I can
pull the blanket off it and drape it down over the opening.
Ostensibly this is because I read before sleeping or upon wak-

ing, and the blanket curtains-off most of the light. They appreciate this gesture, but I know that Pop is a raucous snorer and that the blanket deadens the sound and allows me to sleep.

"Going to eat?" I ask Wade as I pass by his room on the way to the galley. The night crane operator is obliged to sleep with his crew.

"I guess," he says. Wade is still a little pale from the boat ride. He is sitting on the edge of his bunk smoking a cigarette. This is allegedly his last hitch on 45. He is allegedly heading overseas to Brazil in a few days, all "allegedly" because for the past four months he has been promised an overseas transfer. Every crew change he says "good-bye" to everyone, and a week later he is back again. This time, he says, he's certain. He has his passport, his papers. He just has to get a couple of shots and "Brazil, here ah come!" He takes a long drag and lets the smoke out slowly, staring at his feet. "I'll be down later." Why he's here at all, I'm not sure. All this travel for one tour doesn't make sense. "They asked me to . . . 'til the new guy come," Wade explains.

I continue forward to the door leading to the passageway which connects the port and starboard quarters. I exchange the smell of the waste treatment plant for the din of the four diesel generators. The door to the engine room is over one hundred feet away but it is open, and the corridor, with both ends closed, becomes an echo chamber. Immediately to my left is the thyristor room (called in earnest by one roustabout no longer on the rig, the "thyroid room"). This is where the DC current generated on the rig is converted to AC. Jimmy, the electrician, calls it the nerve center of the rig. A deep hum penetrates the heavy door into the corridor.

Between the hum and the din, walking the corridor is akin to running the gauntlet. You almost count the steps until you are through the far door and into the television room, which is empty now. There is a large color television, which doesn't work, in the far corner, a bulletin board with yellowing notices along the far wall, and plastic covered easy chairs in a semicircle in the middle of the room, facing the television. There are a couple more easy chairs and a long, plastic couch along the near wall. To the right is a closed door with a sign that says, "Reading Room." The builders could not very well have named it, "Bou-Ré Room," which it really is. The room looks unnaturally clean. Even the ashtrays are empty. The galley hands must finally have discovered it.

The door to the port corridor is straight ahead. A door to the galley is to the left. I open it and go in.

* * *

Steak for supper. Wednesday and Sunday are steak nights. Friday is shrimp night. The rest of the time it's spare ribs, veal cutlets, pork chops, Swedish meatballs, spaghetti, or fried chicken, all in prodigious quantity. The quality, however, depends on who is cooking and how bad a deal Reliance Marine Catering of New Orleans was able to make when purchasing the food for the week; in other words, the quality is uneven with a tendency toward the dubious.

It is with a blend of hope and hesitation that I walk down between the two long tables toward the counter in front of the kitchen. I love steak, but I've learned not to anticipate too much when it comes to Jackson's steak. As I pass by the port table I glance down at the plates of the men who are already carving theirs. The steaks are large which bodes badly. I look at Bear's plate. He is the best test, for Bear will put up with a lot to clean his plate. Bear is struggling with his knife. Maybe he has a dull knife.

I know better than to ask aloud how it is. Jackson is very sensitive when it comes to his cooking. He has also explained Reliance Marine Catering's purchasing policy which is consistent with its hiring policy—get anything you can for the least amount possible and make it do. The food will be thrown away and forgotten, the galley hands will quit and go away.

So, I'm faced with one of those crucial decisions that I alone can make. Good steak should be rare, but what about tough steak? Do you let Jackson cook it black and hope that the gristle wears out and breaks down so you can cut it—but lose the juice and the taste—or do you go for the possible taste and subject yourself to the exercise of masticating it? As I stand in line, I listen to what those in front of me have decided. There are four men. Three ask for medium well-done, the fourth, Grady, the toolpusher, the saviest man on the rig, says, "medium rare."

"How do you want it, pardner?" asks Jackson.

I sigh. "Medium rare," I answer, then whisper to Grady, "I'm holding you accountable." Grady gives one of his few smiles. Once on the rig, Grady appears perpetually grouchy. He says he is not, that he is just as happy out here as he is back home in Columbia, Mississippi, tending to his small string of Tennessee Walkers. I've seen him both places, but I'd never call him a liar.

In the same way, I'd never tell Jackson his food runs something short of Gallatoire's. I once suggested the reason the steak was so tough was that Reliance Marine Catering made the steers swim out to the rig to save on shipping costs. He didn't smile, but there was a general nodding agreement around the table.

"There you go, pardner," and Jackson hands me a plate with the steak smack across the middle. I thank him, but unlike the other men, hesitate to call him "Jackson." I've asked him if he minded being called Jackson. "It's gotta be better than Wallace," he answered. When I first met him months ago, he was only a cook, a job he took with the proper grain of salt. But Jackson has endured where others have not, and now he is the steward on board for Reliance Marine Catering. With his increase in status and responsibility, he has decided to adopt a different tone, to provide a good example for the help. Consequently, he has lost some of his sense of humor, an absence which could also be attributed to the long hitches he puts in. While other cooks spend two weeks on the rig and one week off, Jackson has been known to work three weeks to a month straight before taking a week off. He says he does this for his family back in Gretna, Louisiana.

I add some sauteed potatoes, corn, and smothered onions to the plate. There is a wide selection in warming pans on the counter, from a pot of gumbo to finely chopped kale. There is even some warmed-over fried chicken of indeterminate vintage. I go to the salad table over in the corner next to the kitchen door and heap a bowl of greens and sliced tomatoes. While there I check the desserts: coconut and strawberry pies, gingerbread, and banana pudding. While I'm standing there, one of the galley hands comes out with two one-gallon containers of ice cream. I'm tempted but decide to wait until the first break for dessert.

There are two long tables in the galley, one to port, one to starboard, with room enough at each for twelve men. The tables have light brown formica tops and four rows of six wooden-backed stools which are permanently fixed to the floor and swivel. On top of each table are two small trays holding various condiments. They reflect the southern taste: Tabasco sauce, pick-a-peppa, catsup, Worcestershire sauce, miniature pickles, various kinds of salad dressing, and a jar of toothpicks. There are also salt and pepper shakers nearby and a paper napkin container.

The two tables are identical in every way except for an invisible line between them which is rarely crossed. Galley hands eat at the starboard table; deck hands and the rig bosses always eat to port. Once in a while when there is no room at the port table, a deck hand will eat to starboard. Even then he does not sit at the kitchen end, but goes up as close to the television room as possible. Divers usually eat to starboard, but away from the galley hands. They have been known to cross over. Special crews like Frank's casing crew or Schlumberger's logging crew also eat to starboard. They come as a crew and stick together. It is almost as though they are not part of the rig when they are out. They don't mingle with the rig crew. They come and go in a couple of days.

And the blacks eat to starboard until they feel sufficiently at ease to move over. When Freeman arrives tomorrow, he will eat all his meals away from us. He will also take his breaks over there. Weeks later he will tentatively join us to port, mostly for breaks, not at meals. Jackson, the steward, who is also black, rarely eats to port although he has been on the rig almost as long as anyone on board and is, in a way, a rig officer. He will sit to starboard and keep up a running conversation with those to port, but he won't come over and sit. Big Daddy and his fellow roughneck, Terry, both of whom are black, always sit to port. There is no one big enough to say they can't. They have also been on the rig long enough to be just roughnecks, not black roughnecks.

This separation is matter-of-fact. No deck hand has ever gotten up from his place because a galley hand or one of the casing crew or a black has sat down beside him.

I sit down next to Dave, the electrician, and across from Otis, the mechanic. Actually you don't just sit down on these stools, you ease into them, particularly if there is someone already on the next seat. The stools swivel at the slightest touch, and, if you are not expecting it, they will turn 180 degrees and you will find yourself lowering down on the back. Worse yet, they are set too close together so that a swiveling stool will often catch the elbow of your neighbor and, if he is left-handed, can catapult a fork full of food off the plate and onto the table.

Dave has already changed into his gray coveralls, his black baseball hat tilted back off his forehead. Dave is always genial. He is rarely without a smile. He is the kind of person who wants to go out of his way to help anyone. A former "lifer" electrician in the navy, Dave is retired at the age of forty-three and has started up a new career in the oil fields.

"How'ya doin'?" Dave asks, making room. "Good week?"

"Good enough," I answer. "How was the fishing?" Dave is an inveterate fresh water fisherman. He and Hank, the welder, spend most of their weeks off on the lakes and rivers outside of Alexandria, Louisiana, where they both live.

"Incredible. Never seen so many bass—"

"You ain't seen nothing till you see 'em back home," Otis interrupts. "Why I just sat on the bank of this little river that runs through this piece of property . . ." And Otis is off with the lure. For the next ten minutes Otis regales us, cast by cast, with the luck he had fishing the river that runs through the piece of property he and his mother are trying to buy back home in Brookhaven, Mississippi. We let him carry on, aware that once started, Otis' mouth is in perpetual motion. Dave is being polite, and I am amazed: Otis is maintaining his line of patter while hacking brutally at his steak, sawing off a piece, getting it into his mouth, apparently chewing it—and not missing a word.

Otis is one of those characters who, if a crew does not have one, it should. His real name is Alton Otis McCall. Or maybe it's Otis Alton McCall. Some call him Alton, most call him Otis. Everyone likes him because it would be almost unAmerican to dislike him. He is twenty-five years old, unmarried, and frequently on the verge of asking some girl to marry him. A Vietnam veteran, he lives at home with his mother in a house he bought for her. He is about five feet, seven inches tall, as skinny and taut as an index finger. His head is small, his face ancient and hairless, his nose protruding and pink from constant dripping and blowing, and his sparse crop of fine, light brown, curly hair pulled back from a receding hairline.

Otis is the most worrying and happiest man in the crew. He loves his job obsessively. Ever since he left the navy, Otis had worked as a roustabout on oil rigs. He has been with 45 since it was launched in 1975. He is its unofficial historian. Few things have happened on it, few men have worked there that Otis doesn't remember. There are those who say his memory is a little vague, that he will foresake an element of truth for the good of his stories, but no one denies that his reservoir of 45 trivia is overflowing.

Hardworking and eager as he now is, Otis was once not entirely happy. More than anything else he aspired to become a mechanic. On his own time he studied the machinery. He learned how every mechanical device on the rig worked—an awesome task since the rig is, in the final analysis, the sum of a large collection of machines. And he pestered the toolpushers as only Otis could, politically stopping just short of being tossed in the burn basket and dropped overboard.

Then, a few months ago, he got his chance. A mechanic quit. Otis was given the job. Now he struts like a dog with a bone, a greasy rag hanging from his back pocket like a wagging tail. He is busy, perpetually scurrying about the rig searching for things to fix, always on the move . . . unless he is in the galley, where, with a cup of coffee and a cigarette, he can be detoured into a story about whatever it might be that is passing through his head at the time.

There isn't much other conversation at the table. The day crew has already gone to work for the hour and a half that is left of their tour. The night crew is passing itself through the galley door to port, those who ate first leaving to dress, those having dressed coming in to eat. The effect of the crew boat ride has worn off to be replaced by a general weariness preceding the long twelve hours about to begin. Grady has given up on his medium rare steak, has turned sideways in his stool and is staring at the blackboard along the wall while picking his teeth. Carney Ewing, the barge engineer, is staring from across the table, past Grady's shoulder. He, too, is picking his teeth with the preoccupation of a cat cleaning himself. Clyde, the driller, has a cup of coffee between his hands. He is staring absently at the condiments in front of him while beside him Hippy is hunched over his plate, working determinedly at the steak, and once in a while turning his face toward Clyde and almost whispering to him some classified information which undoubtedly has to do with one of Hippy's putative women. Clyde signifies he's heard by lifting and dropping his chin.

By now it's almost five o'clock. Time to get dressed. I snuff out my cigarette and swing gently out of my seat, careful not to bump into Pop, who has just begun to wrestle with his steak beside me. I head diagonally across the galley to the garbage pail in the far corner by the kitchen door. As I pass by Grady, I say,

"You blew it."

"That was an old hoss, wasn't it?" Grady answers. Most of his steak is still on the plate.

I estimate from the pile in the garbage pail there is enough flesh left on the bones to put the steer back together again. So much for Jackson's supper. I'm already looking forward to his midnight lunch.

I go straight down to the locker room and get my work clothes out. They are "oil rig clean" which means that any place else they would be considered filthy. The oil and the mud has so penetrated the fibers that no amount of washing will ever make them presentable. It's all right. An hour from now they'll be even more grimey.

I go back to the room and stow my gear. Pop has already established his territory. The only chair in the room is by the head of his bed. He has found an ashtray and put it on the seat. There are two squashed cigarette butts in it. His yellow hard hat is square in the middle of the bed, beside it a pair of work gloves on which he has already penned in "H. J. Carson" across the backs. He is very possessive about his gloves. They cost one dollar a pair and will last three days at the most. With all the discarded gloves lying about the rig, it is possible to work for months without ever buying any. Still, it is uplifting to set off to work the first tour with a pair of spanking new gloves hanging out of your back pocket.

I take off my wedding ring and stick it safely in the back of the top shelf of my room locker. One of the ways you can tell who works on deck and who doesn't is by whether he is wearing a ring. There are stories about those who neglected to remove their rings: how a frayed wire from an old cable cut through a glove, caught under a ring, and snapped the finger. General

Safety Rule No. 16 states: *Finger rings are not to be worn during work.*

I also stick my wallet back with the ring. It has about sixty dollars in it. Some of the men lock their wallets up in the main locker rooms. They argue that there is no sense in providing temptation. The counter argument is that the rig is too small an island. No one is going to steal anything. The rules about stealing are harsh and absolute. The slightest suspicion and the suspect is "run off" the rig without a defense. The judge and jury is the toolpusher.

I dress and lie down on my bunk. It feels good. I'm ready to go to sleep and Pop walks in, hauls a blue bowling bag out of his locker, sits down on the edge of his bunk, and opens it. This is his medicine chest.

"You want one of these?" he asks, holding up one of the many brown bottles he keeps in the bag. "It'll pick you up for the night."

I thank him but "no," and watch as he swallows an assortment of brown and yellow and red pills. Pop has a litany that goes with his pills. All they are are various vitamins but as far as he is concerned, they are his fountain of youth. The selection for the day depends on how he is feeling. Time has taken its toll on Pop's body, both inside and out, his hands especially. They are scarred, the fingers crooked from dislocations and breaks. He is deaf in one ear; his "smeller" is burned out, but he does have most of his teeth and a good part of his hair.

"Time we got up there, I guess," I say reluctantly, swinging off my bunk. I reach over the locker and take down my hard hat and my new pair of gloves.

"We can let 'em wait a few minutes," Pop says, but he's already standing, too.

"Don't forget your Copenhagen," I remind him, as if Pop would ever go to work without his snuff.

"Which reminds me," Pop says, and takes out his snuff can and serenely places a pinch behind his lower lip. "Now I'm ready. Let's go see what they've got in mind for us tonight."

"Your hat, Pop."

"Unh," he grunts and fetches the hat off the bed. We head down the corridor, retracing our steps through the mud room, up the stairs, and out onto the mist-slick deck.

*   *   *

The changing of the shifts, or the tours, is curiously similar to the changing of the crews. Depending on how hard the twelve hours have been, the change is made with either a wave of the hand or the exchange of a few desultory words. If you are late in arriving, the off-going crew will say something like, "We'll pay you back in the morning," meaning they won't appear on deck a minute before five-thirty the next morning. If you are early, you are careful to point out the fact, saying, for instance, "See you boys bright and early," to which they will answer as wittily, "Don't count on it."

But there is an unwritten rule which says that while you are not expected to be on deck before five-thirty, it is best to be at least five to ten minutes early. The total time on deck is still twelve hours, but psychologically that early relief gives the impression of getting a break. Five minutes at the end of a twelve-hour day seems like half an hour.

Byron and Dwayne, the "kid" roustabouts from Mississippi, are standing together behind the red Otis tank. As usual Nick, the loner from Texas, is by himself near the lifeboat. The two crane operators, Wade and Jake, are walking toward us from the port crane, discussing what the day crew has done and what has been left over for the night crew. They break apart. Jake strides past without a word to us.

"Let's go down, men," he says to his crew. They follow him down the nearest stairwell. We won't see them again until the next morning. If the week goes according to schedule and neither crew is held over or called up for overtime, this brief exchange each switch will be the extent of our communication.

"See you in the morning," I say to Nick as we pass.

"Have a good night," he answers. Which is a mouthful for

him. I notice he has shed his old gray sweatshirt and blue jeans for new brown coveralls with NICK sewn onto the left breast pocket. I wonder whether this means Nick has made a decision. He is three years into the University of Texas, majoring in political science. He left for a semester to earn the rest of his tuition. He has been on the rig for three months and has been wondering whether he would go back to college. He says he likes the work; he also likes the money.

Pop and I are left standing alone. Without a word, Wade has turned around and disappeared into the day office at the bow of the rig. The day office is a gray, two-storied shack. The top story is the radio room. It isn't used much. Cups of coffee often sit moulding on the table. All lengths of antennae stick out from the flat roof. The first story is the toolpusher's office. There is a desk in the left-hand corner. It is usually in a state of disarray. Most of the papers on it appear to lack purpose, except to give the impression of an office. The situation of the desk is what counts. From his chair, the toolpusher can survey a large part of the deck. Experienced roustabouts know this; new roustabouts learn quickly.

The day office is also a quasi lounge for the bosses. If you can't find a crane operator or a driller or the barge engineer, you go to the day office. If they are not there, then try ballast control. The bosses can be a little more conspicuous when they have nothing to do; roustabouts and roughnecks have to be more clever—next to being careless and endangering someone else's life, the worst crime a roughneck and a roustabout can commit is being caught sitting down.

The day office is likewise the rig's library. On the two shelves over a long table behind the toolpusher's desk is an assortment of technical manuals describing the workings of various pieces of equipment on the rig. These are clearly for reference. There is a thin tell-tale layer of dust on the tops. Better read is the newspaper on the couch where the night toolpusher catches his midnight catnap, or the paperback, dogeared and open, on the desk, or the *Penthouse* and coverless *Playboy* in the desk drawer. More issues of the latter two can be found in the drawers at ballast control. Some are secreted in the tool-pusher's private office adjoining his sleeping quarters down below. There are issues dating back years.

The day office is considered to be on the bow which puts the heliport to stern. This takes some getting used to because the rig is nearly four square, being two hundred and eighty-eight feet long and two hundred and sixteen feet wide and the designation of any one part as a bow or a stern at first appears arbitrary. Initially I suspected the theft of such nautical terms was inspired by the need for a port and a starboard, it being awkward to order someone to fetch a set of tongs from the left storage room when, without a point of reference, left could be just about anywhere.

The fact is, despite its unseaworthy appearance, *Gulf Star 45* is a "floater," a self-propelled semisubmersible rig, and therefore a U.S. registered vessel. At the foot of each corner leg is a two-thousand-horsepower thruster used mainly for holding the rig on location but capable of moving it forward at around four knots per hour. As it sits firmly anchored in the southwest corner of Block 6—, it gives the impression of iceberg perma-nence. Its six gigantic legs seem to be mere tips of shafts which must extend at least to the bottom of the Gulf. On the contrary, un-iceberg like, the bulk of the rig is what you see—seventy feet of leg to the rig deck, another thirty feet to the drill floor above the middle of the deck, and a final ninety feet to the crown of the derrick which peaks above the drill floor. The legs extend only thirty feet below the water where they fasten onto a pair of huge pontoons. Crisscrossing between the pontoons are equally voluminous cylinders sectioned into ballast tanks. These are filled and emptied with sea water to raise and lower the rig, and, as importantly, to keep it in ballast. Ballasting is the sole job of the barge engineer and his assistants, the watchstanders.

"What're we supposed to do?" I ask Pop, the senior mem-ber of our depleted crew.

"Fuck the dog 'til Wade tells us something, I guess," he answers.

Waiting for Wade to give instructions is like waiting for

lightning to strike you. Wade is a very good crane operator, but a lackluster boss thoroughly trained in the "hurry up and wait" school. He is also the kind of person who knows what he wants you to do and forgets to tell you until you haven't done it, then he tends to lose his temper. I've often wondered whether this trait was only a lapse in memory or was a design brought about by Wade's shortness and the bulk of most roustabouts. I do know that the turnover rate amongst his crew has been high, sufficiently so, in fact, that six months ago he was transferred from the rig to learn a little humility. Since he's been back, he has been easier to live with, but now that he is certain he is going overseas, a certain arrogance has returned.

"We could start scrubbing deck," I suggest.

"We'd better wait," Pop insists. We do. There is no sense scrubbing the deck too soon. Jobs like that are for times when there really isn't anything to do.

"Well, I guess we could put some of these slings away," Pop says. Pop doesn't like standing around. He also has a fetish about stray slings. They spoil his sense of order. He reasons that if they are all in one place, when you need one in a hurry, you can find one. He says this is more efficient than searching all over a deck which is practically as long and wide as a football field is long and which is littered with equipment which, like the slings, have been dropped at the nearest convenient spot. General Safety Rule No. 1 states: *Good housekeeping is a must.*

Pop is right, of course. At our feet are at least four separate slings plus an array of shackles. He grabs the ring at the end of a long double sling used for lifting very heavy equipment. I pick up the hooks at the two ends. We shoulder it and pick our way down the deck, stepping over hoses and pails, past the stack of risers in front of the shale shaker room, by the starboard crane. We walk carefully. The deck is slippery from the day's rain and the present mist, from the hydrolic fluid which has leaked from the crane, and from the drill mud which has spilled from the drill platform to our right and up thirty feet. We make slow progress. Walking on the rig is dangerous. The stretch of wire between us is slack and drags on the iron floor and catches on corners and pipes which stick out, stopping us short. The rig itself is rocking in the twelve-foot seas breaking against the legs below. The rocking is subtle, not calculable as it would be on the deck of a boat, just enough to keep you off balance.

We make it to the catwalk on which lie some sticks of drill pipe ready to be dragged up the V-door to the drill floor. The catwalk is about seventy feet long and stands ten feet off the rig floor, supported by heavy pipe. It has a yellow railing on each side and thick cypress planking on the walk. The railing is bent and dented from being hit so often with swinging pipe. The V-door is also made of cypress and rises at an angle to an open space in the iron wall about the drill floor. A heavy tattered canvas flap is drawn across the open space to keep the wind out while the roughnecks are working.

Under the catwalk stands a row of blue fifty-five-gallon drums containing diesel fuel for the rig generators directly below deck. And over the barrels are iron pins to hold the slings. Pop heaves his end up and catches it on a pin; I heave mine. One down. We turn around and make our way back along the deck to fetch the rest. We don't hurry. It's going to be a long night.

Still no sign of Wade so we fetch the remaining slings and haul them under the catwalk. Then we go back for the shackles. Along the way I notice a sledge hammer lying under the risers and bring that along, too. Pop picks up a pair of squeegees and a worn-down, long handle scrub brush which he sticks by the crane.

We are being good roustabouts, good dirty-work men. It doesn't take much to be a roustabout. "Just keep your head down and your ass up and push, you'll do all right," an old service hand once told me. It is menial work of the rawest sort. A roustabout is called on to do just about everything except something important, and sometimes, when there is no one else available, he gets to do that, too. He is at the beck and call of everyone, from the lowest roughneck to the toolpusher. He is prey to the whims, caprices, absurdities, angers, frustrations, and humors of anyone above him. He does not, however, have to answer to a galley hand, although he may be ordered to help one. This is not considered abasement but rather kindness to a lower order. Such a service is rendered infrequently and usually involves stowing heavy cartons of food stuffs which the roustabouts have already unloaded from a supply boat.

There is an advantage to being a roustabout. Money. It is one of the few jobs at which someone with a strong back, a high tolerance for pain and subordination, few skills, and little education can make $250 a week plus room and board. But more importantly, it is the only way to make roughneck, and that means more money and a step closer to driller, which is more money still, and finally to toolpusher—the top money, power, and prestige. The fact is, in order to do anything on an oil rig, you have to go roustabout first. If you can do that, as a driller once said, "you can go anywhere in the oil field—roughneck, crane operator, ballast control." The reason is simple: after you have been a roustabout for a while, you have done everything, or at least seen how it is done. There is an old oil field adage: "If you can't do it, you can't stay." Roustabouting is how you learn to do it. Those who can't learn or refuse to, don't stay. No one wants to be a roustabout for long.

"What y'all doin'?"

It's Wade, back from the day office. It's also not really a question. It's almost an accusation. He's standing by the port crane. Short, broad shouldered, stubby fists sticking out from an old thermal undershirt pulled over a thermal vest buttoned over an old gray T-shirt, he has on his pinstriped overalls with the ragged suspenders which have been mended over and over again for the last time. When he leaves for Brazil, the overalls go into the wastebasket, Wade has declared. He has a cigarette hanging out the corner of his mouth. The two small eyes high up in his round face are red and watering. Wade always seems to be crying. Actually he is prone to colds.

"We were just putting away the slings there," Pop answers obsequiously. Pop tends to develop a flaccid backbone in the presence of bosses. He stops just short of "sir" to Wade, who, at twenty-five, is half his age.

I notice a paperback sticking out of one of Wade's back pockets. I know what Wade has been doing while we've been working.

"Waiting for your orders," I answer.

"Well, y'all get yourself a double sling and take it over to the burn basket. Real quick now. . . . Pop, you get yourself a bucket of diesel fuel."

And Wade turns around and goes back toward the day office.

There are two burn baskets, one to port, the other to starboard, both to stern. They are filled with garbage and trash from the galley, empty chemical bags from the mud room, pieces of frayed rope, broken two-by-fours, and anything else someone imagines can be burned and has not thrown overboard instead. The baskets themselves are large, open boxes framed by welded iron bars with thick iron mesh around the sides and across the bottom.

Burning trash is always done at night because it is illegal. The burning itself is permissible; it is the dunking of the basket in the Gulf to let the debris float away that is not. The theory is that by the time the U.S. Coast Guard spots the debris it will have drifted far enough from the rig.

We presume Wade means the port burn basket, since he was standing by the port crane when he gave the order. Because of Pop's insistence on efficiency, we know where the double sling is, where the shackles are. It is merely a question of now completely retracing our steps and taking the sling back where we found it. This is one of those frequent instances on an oil rig when inefficiency would have been the quickest route.

While Pop is draining some diesel fuel out of the port crane into a mopping bucket, I shackle the sling to the burn basket, which is hanging on two iron pegs off the side of the rig near the lifeboat.

We are a smooth team tonight. We don't need Freeman, we haven't missed Bear. We may be tired but our heads are down, our asses are up, and we are pushing together. Only we seem to be missing our driver again.

"Where the hell is Wade?" I ask.

"Taking a shit, I guess," says Pop.

"Shooting the shit is closer," I suggest.

Wade doesn't have to go to the bathroom, at least not the way Pop means it. "Taking a shit" is a popular work dodge. That and eating are the only times we are allowed to sit down during the twelve-hour tour. There are those who seem to be called more often than others. Pop rarely feels the urge more than twice a tour, usually during the back half; I try to follow suit, declaring the need sometime after both breaks. Freeman tends to beat a fairly steady path to the bathroom—the one farthest from wherever he is at that point. Freeman also cheats. He goes below to take a leak, too, whereas most of us are not so brazen and stand to some leeward point and go off the edge of the rig.

So we wait for Wade.

"You heard about the guy with the fastest bull in the state of Mississippi, didn't you?" asks Pop.

"No, Pop," I answer in dread. Pop's stories can blister paint.

"Well, seems there was this farmer who had this good looking bull and one day there's a knock on his door and there's a man who wants to buy the bull, only it's got to be a fast bull. The old farmer looks at the man and tells him,

" 'See that bull over in that field yonder? See those train tracks running by the field? At five o'clock, the train to Jackson comes roaring by, and that bull takes off along side it and beats that train to the other end of the pasture. Done it every day for years.' "

"That's fast enough for the man. 'How much you want for that bull?' 'Five hundred dollars, I'd let you have him,' the old farmer says. The man says he'll think about it.

"Next day there's a knock on the door. The farmer looks out the window and sees this guy standing there. He turns to his daughter and tells her to get rid of him, tell the guy anything. She answers the door. The guy says he's come to buy the bull.

" 'Well, my Daddy ain't here right now. He's gone up to Jackson, then he's going up to Nashville 'fore he comes home,' she says."

" 'When you expect him back?' "

" 'Bout four o'clock,' she says."

" 'What he take—a plane?' "

" 'Hell, no,' she says. 'He took the bull.' "

I give Pop a straight look as though I don't get it.

"That bull couldn't run. The old farmer knew he'd—"

"I know, Pop. I know."

"Oh, . . ." says Pop. There's a moment of silence, then we both laugh. "You know what the finest thing in the world is?"

"Yes, I do." I do know. It is one of Pop's favorite one liners and he is just old enough to be a repeater, so I've heard it at least a dozen times. It's foul, not very funny, and also doesn't make much sense.

I can tell Pop is warming up so to head him off I am about to tell my story about the three bulls when there is an insistent blowing of a horn from the opposite side of the rig. It has to be Wade on the other crane. Only the cranes have horns.

"What's he want now?" Pop asks.

"Pipe on the catwalk? I don't know."

We set off for the other side of the rig, Pop in the lead. He has a funny shuffle, as though he were sloshing through a puddle. With his rounded shoulders and old bagging pants he reminds me of Emmet Kelly, the clown.

"Where y'all been?"

In the steady roar of the rig it is hard to hear anyone more than ten feet away. With the big Caterpillar in the crane racing it is nearly impossible. To the Yankee ear the Mississippi dialect is often indecipherable under the most tranquil conditions. I know what Wade has said because I know Wade well enough to anticipate him and I can see from the way he is sitting in the crane and looking at us what is on his mind. For Pop, Wade's words are a mumble. He turns his good ear toward the crane:

"What's that?"

Wade repeats himself.

"What's he saying?" Pop asks me. I shrug. I'm in no mood to deal with Wade. We have this one tour left together. Tomorrow night we'll have a new crane operator. If Wade wants to be feisty, let Pop handle him. He's had more experience with bosses.

Pop shuffles up beneath the crane. Wade has pulled the window down and is leaning cross-armed on it, glaring down, the ever present cigarette drooping from the corner of his mouth. I lean up against one of the spare, thirty-thousand-pound U.S. Navy lightweight anchors sitting on the deck. It is still caked with mud from the last hole we drilled. It smells of low tide, pungent but one of the few signs that we are actually at sea and that the rig is also a floating vessel. Otherwise, from the sounds and the smells and the junk around the deck, we could as easily be back in Sabine Pass.

Pop trudges back and walks past me. I join in step with him.

"He wants to burn this one first," Pop says. Once I would have asked why. Instead I go with Pop to shift the double sling from the port basket, aware that once done we'll be taking it back to port. The only thing we will have gained is more time wasted. It is already nearly nine o'clock, and we have succeeded in accomplishing almost nothing. It appears that burning the trash may be the high point of the first tour.

There are few things aesthetic about work on oil rigs. Burning trash is one of them, although there is nothing charming about actually setting the trash ablaze. That part is maddening, especially when there is a wind blowing as there is tonight. The trash itself is wet. Pop pours half a bucket of diesel fuel over it. I dig through the garbage for a dry wad of paper. Pop soaks the paper in diesel and I try to set it afire. The matches blow out. Pop suggests I try two at a time. Two stay lit; the wad of paper flames up. Pop quickly drops it on top of the trash. It sputters and goes out.

"Go down below and get some napkins," Wade yells.

"Get some more matches," Pop suggests.

I run below to the galley and grab a handful of napkins from the table, stop at the milk dispenser to pick up a couple of boxes of Reliance Marine Catering box matches, and run back. This is an instance when you must hurry. Although nothing much has preceded the trash burning, and nothing much may occur afterwards, burning the trash is an act as important at this point in time as quelling a blowout.

Pop has cleverly constructed a windbreak out of rain-wet and diesel-soaked boxes. He has piled an array of potentially burnable items inside the windbreak. He is all business when I arrive. Wade is still leaning on the crane's window sill, smoking and staring.

We douse the paper napkins, set them thoroughly on fire before dropping them carefully into Pop's nest. The fire begins to catch. Pop steps back and signals to Wade to hoist the basket up. He is full of confidence. The crane roars to life, a blast of exhaust streams from the stack. The sling goes taut. The trash is catching, flames are beginning to swirl in the wind. The basket comes neatly off the pins and begins to swing . . . and catches on the rail.

"Push it," Pop yells. I shove, and it becomes more firmly wedged.

"Get the fucking thing loose," yells Wade. Pop and I both shove, to no avail.

I motion to take the basket up a touch higher. Pop motions to lower it.

"Get it up," I yell.

"Drop the sonuvabitch," Pop yells. We are still both shoving. The smoke and flames are blowing into our faces.

Suddenly Pop jumps up on the rail. "Hold onto me," he yells.

"Get off that thing, Pop," I yell back, but he is already

jumping up and down on the near corner of the basket which, without warning, breaks loose, tumbling Pop back onto the rail and down on me.

"You're a crazy bastard," I tell him as I look down between the edge of the rig and the corner of the basket to the rolling seas seventy feet below us. Typical, I think to myself. At one moment we're standing around doing absolutely nothing, the next minute we are risking our lives, and then we are back idling away again, this time watching the burn basket descend to the seas where Wade holds it. The winds have fanned the flames into a roar. Sparks race after one another like tracer bullets. All three of us stand there absorbed in the celebration going on so many feet below. The breaking waves glisten in the light of the fire. The rampage of the sea is magnified. The scene is curiously arresting.

"Let's go get some coffee," Wade yells from his seat in the crane.

"Break time," Pop says.

II

I hang my white hard hat next to Pop's yellow one on the hooks outside the galley door. You can tell who is inside by the hats outside. There's Clyde, the driller's, red one with the ARMCO sticker on the side and Catfish's handpainted hat of many colors. For many of the oil hands, collecting oil field stickers for their hard hats has the same enchantment that collecting baseball cards has for children. It is illustrative that Clyde should have the only red hat on the rig and that it should display a number of stickers whereas Catfish, who has been roughnecking for almost as many years as Clyde, should not have any. Dan Walton, the night toolpusher, has the numbers of all the rigs he has worked stuck to his. He has only one sticker, that of the Gulf Star Drilling Company, the only company he has ever worked for. Dan is considered a man on the rise. Interestingly, Dan has a hard hat while Grady, the day toolpusher, his immediate boss, doesn't. Dan spends more of his time on deck; Grady does his work in the day office watching out the window, letting the news come to him. Wade also does not have a hard hat whereas Jake Pitre, the other crane operator, does. Wade's deck time is limited to sitting in the crane, Jake has been known to work with his men. Jake is still new to his job but he has been seen less and less with his hard hat. He is learning.

"Si' down, Perfessor," Catfish says. There is an empty seat across the table from him. I think I will, but I give him a quick glance to assay his mood. Catfish is given up to moods. In the course of a single tour he can run the gamut of emotions, yet never change expression or alter the tone of his words. The only way you can tell how he is feeling is by timing silences between short bursts. The longer the silences, the worse his mood.

Catfish is considerably more bark than bite. Still, he takes getting used to. He is one of those people who long ago reached his level of competence, realized his place, and decided not to seek higher ground. He is thirty-six years old, married with two daughters and a son at home in Camden, Arkansas. Years ago, he chose between the local pulp mills and the oil fields, went where the greater money lay, and is now a roughneck working the back-up tongs which puts him one notch higher than Bear, who handles the lead tongs.

Like "Bear" and "Hippy," "Catfish" is a rig nickname. His wife calls him Quentin. Once Grady, who is called Austin at home, had to telephone him. His wife answered and told Grady there was no one named Catfish at that house. "That's me," came a voice from across the room. "They call you, 'Catfish'?" his wife said . . . and began to laugh.

Just why he's called Catfish is a matter of some disagreement. The southern fresh water catfish is distinguished by a mean appearance marked by pronounced whiskers. Catfish has a beard which is unique in that the hair is quite straight and wants to turn away from rather than toward the face. And he looks mean. But appearance, while perhaps the genesis, is no longer the primary explanation. Hippy says it's because catfish are renowned for their stupidity and "Have you ever tried to talk with Catfish?" I have and I found him anything but stupid. Uncommunicative? Maybe. According to Clayton, the derrick-man and a fellow Arkansan, the catfish makes an eerie grinding sound with its teeth as it expires after being caught. Clayton says that when Catfish sleeps, he makes the same sound. I've never bunked with Catfish, but those who have corroborate Clayton's claim.

All I know about Catfish is that both he and Bear insist on wearing their hard hats backward, that he is a good worker ("you know he ain't goin' to screw up on you"), and that he calls me "Perfessor" because he found out one day I'd been to college and majored in "What! Philosophy? Hey, Perfessor!"

Breaks are often as much of a production as meals. For some, like Wade and Hippy, they are meals. There is usually hot gumbo simmering in the warmer in the kitchen counter. And there are always desserts on the far table by the kitchen door. But breaks tend to center around the counter along the wall separating the galley from the television room. From right to left there is a large ice chest with a rack above holding clean glasses. Next is the milk cooler holding two five-gallon containers of milk, spouts sticking through the handles to lift to release the milk. Then there is a Kool-Aid dispenser, one half for grape or strawberry, the other half for orange or lime. Beside this there are two coffee pots, both yellow, one marked: "Dark," the other "Light." To the uninitiated there appears little difference between the two—until he tries the dark. One cup guarantees a sleepless night. Which may be why there is a stand of paper demi-tasse cups near the bowl of sugar. There is also a pot of iced tea near the coffee.

Finally, there is an old refrigerator containing jars of mayonnaise and mustard, cans of fruit juice, various cold cuts, and loaves of bread. As if this were not enough, in a cardboard box under the coffee pots sit many cans of what Wade calls

"Vy-eeny Weeners" (Vienna hot dogs). These are very popular during breaks. Hippy and Wade will go through two or three of them. Pop is always good for one. Pop eats his spilled out on a plate, mustard on the edge, with a slice of white bread to soften the sting. The others take toothpicks and pluck them from the small tins with a side order of crackers, also found, in cellophane packages in a box, under the counter.

If it is fair to say that the world is divided between those who drive Ford cars and those who don't, it is equally justified to suggest that the crew of *Gulf Star 45* is divided between those who will eat "Vy-eeny Weeners" and those who can't stand the sight of them. The result is one half the crew absorbed in diverse versions of a "Vy-eeny" repast and the other half casting sidelong glances and disparaging remarks. Which simply leads to a heightened creativity on the part of the "Vy-eenies." Entire breaks have been spent with the cons describing in visceral detail what the nearest pro is now doing.

But tonight the men are more interested in hearing about Wade's latest brush with the Louisiana State Police. Per usual it has to do with Wade's speeding, but everybody knows that in Wade's case that relatively minor offense can have ramifications. His nickname is "Weed" and it didn't derive from an ill-kept garden. Wade likes his smoke, and while he swears he doesn't use any on the rig, he states as fervently that he makes up for the abstinence the minute he's on the bank. Which means that he has his road supply in his car immediately handy to him as he is driving. The car could not withstand a search.

". . . I don't know where this guy comes from. I got Bull-rider keeping watch and I don't think I'm going fast. I'm feeling real good, though, so I don't know . . . when he's passing me and he's got that little ol' blue light on. He keeps right on movin' so I guess he's after someone else when he slows down real slow and pulls off to the side of the road.

"Bullrider says to keep it on. Fuckin' Bullrider! I guess maybe I'll stop and see what he wants.

"Out comes this big ol' boy. You know how it was raining

and I can see this guy don't want to be out there and I'm not
going to keep him, I guarantee you.

"He says I'm goin' seventy-five. I couldn't be. I know I
wasn't, but I don't want him around the car. Geezuz, I got a
roach still hot in the ashtray. Bullrider's got his window down
all the way. Rains blowin' in. So I says, "Yez, sir, that's right. I'm
mighty apologetic." And the motherfucker hits me with a one
hundred and twenty-five dollar fine. One hundred and twenty-
five dollars! Right there. I got the money and I know I'm not
goin' to the station with him. I give him the money and he's
gone. I like to tell ya, brother . . ."

"Hell, you say," Clyde says.

"You tell Hippy?" Catfish asks.

"Nah, not yet," Wade laughs. "Ol Hippy. Man, would they
love to get into his car."

"You lucky, babe," says Clyde.

"I guarantee you," Wade agrees.

"What you going' do down in Brazil for twenty-eight days
straight?" Catfish asks. "You goin' be nasty 'for your hitch is
done."

"But I'm goin' to be mellow when I'm home, you can count
on that," Wade says. And turns to Pop and me. "We'd best get
that other burn basket, men."

I can see Pop is beginning to fret. It's ten-thirty. We've
succeeded in setting the port basket ablaze. Normally we would
now stand around, smoke a couple of cigarettes, and watch it
burn out. Depending on the amount of wind this could take
upwards of fifteen to twenty minutes and be construed as con-
structive work, especially on a night when useful work appears
to be at a premium.

But Pop is one of those people who cannot stand around
doing nothing. Regardless of how menial the task, as long as it
keeps him moving, Pop will seek it out. Given such a disposi-
tion, Pop has long since assumed responsibility for washing the
clothes of the off-duty roustabout crew. And typically, he has

developed the simple chore into a production of profound complexity. It is a sign of the relative position of the roustabouts that they are required to wash each others' clothes. However, those are all they have to wash. The galley hands must take care of the roughnecks', welders', mechanics', electricians', and watchstanders', as well as their own clothes. Why they don't handle the roustabouts', too, is unclear. Perhaps at one time the galley hands asserted themselves and pointed out that of all of the deck hands, the roustabouts could best find the time. At any rate, Pop inherited the job to which he has given an importance heretofore unsuspected—and to date unchallenged—by either his fellow roustabouts or the crane operator, who could offer an argument that Pop might be abusing this responsibility. It is accepted that this is Pop's way of "taking a crap."

Now, it would seem to most people that washing clothes would be a modest task, but according to Pop it is anything but. After each tour, the off-going crew returns to its various rooms and undresses. How each gets his filthy clothes into the corridor is a matter of personal preference. Some strip in the corridor and kick their clothes into a pile against the wall; others leave the door open and toss each piece of clothing out into the corridor and when they leave to shower, kick them into a pile; the rest pile the clothes first in the rooms and move the pile to the corridor later. Sometimes, there are separate piles outside various rooms; other times, all the clothes are bunched together in a heap. Pop says the heaps are the most difficult to handle, especially if there are three sets of clothing involved. Getting them to the washer is not the problem, according to Pop, separating them afterwards is. For Pop is meticulous, as is Nick, who does it for us. They both insist on making separate clean piles. To those of us getting up groggy, not having to unscramble clothes is appreciated.

There are two laundry rooms on the rig, both aft; one to port, the other to starboard. In each there are two washers and two driers, a long sorting table, and a barrel of laundry soap.

Each washer and drier is designed to take half an hour to complete a cycle, so that each load should take an hour. Put another way, as there are two rooms of roustabouts, there are two heaps of clothes and given the four sets of washers and driers Pop should be able to accomplish his task in an hour. Still, this means going below three times—to collect and wash, to dry, and to separate and return to the proper doors.

But, more often than not, one or two of the eight machines involved isn't working, which forces Pop to use both laundry rooms at the same time or to double his trips below in order to juggle the loads among the working machines. And then there is the instance of the galley hand, who seeing two full driers at the end of their cycles and needing them both for his work, throws all Pop's clothes together on the table; or thinking that two driers with light loads is inefficient, lumps all Pop's together so he can use the other. Any one or a combination of such events produces for Pop a calamity capable of rendering him dour for the remainder of the tour. Since one thing Pop cannot stand is a missing sock or glove from his own pile, he is excessively careful about the socks and gloves of others'.

Consequently, in order to expedite the most worry-free wash Pop likes to get to the job early. He lives in Pop-like fear that we will be suddenly given a job like unloading a supply boat which can take up most of the tour and the clothes will not be dry for the other crew.

So Pop is fretting. He had counted on using the time we would spend watching the burn basket to get his laundry started and now Wade has decided to abandon the port crane and its flaming load to move the pallets of chemicals off the starboard deck and down into the mud room. In the light rain and the gusty wind this is not going to be a simple job. It will probably take us until lunch. I can almost see visions of wet laundry flapping in Pop's imagination.

"What's the matter, Pop?" asks Wade. He can see the same thing.

"I got all that laundry to start yet," Pop says anxiously.

"Oh, for Christ's sake, Pop, you got all night," Wade says, a flicker of a smile in the corner of his mouth. "Everything's going to be all right."

"Well, I don't know. They still ain't fixed that starboard drier and them others you got to go through twice sometimes before you get things dry."

"Jesus, old man," says Wade. "Them other boys'll live. Let's go get the chemicals before they get ruined."

The way Wade squares his shoulders and hustles across the heliport towards the starboard crane you'd think the saving of the chemicals constituted a holy mission, the urgency of which escapes me. There they are: six groups, two pallets high, ten layers to a pallet, five sacks to a layer. Some pallets are tightly wrapped in heavy-mil polyethelyne; the others are open to the weather. Altogether there are probably three to four thousand dollars in chemicals there. Which should make them worth getting below quickly.

But when did they arrive? They didn't come out with us; no supply boat has been here since we arrived. All the supply boats I know of I saw loading at Sabine Pass. Answer—at least two or three days ago. Why then weren't the chemicals stowed after they were unloaded? There have been at least three, maybe four tour changes since they arrived. Our own day roustabouts were on deck for an hour and a half. Answer—it is an unpleasant job, one to be bumped down the line until the doing of it becomes unavoidable.

As I reach for the bars we use to lift the pallets, I try to stifle a yawn and succeed in tripping over a steel plate one of the welders has cut and simply left on the deck.

"Wake up!" Pop barks. "Pay attention."

"Sorry," I mumble and fake a couple of slaps to the side of my face.

I've suddenly developed a case of the Monday Morning Blahs. Pop has them, too. We all do. First day out. But this isn't the time to indulge them. Stowing chemicals isn't difficult, but it's dangerous. Everything you do on an oil rig is. Men are forever being injured. Most of the time the injuries are minor, though painful: smashed fingers, hands, toes, bruised shins. Two hitches ago, a diver slipped on the deck and busted his leg. Last hitch, Big Daddy got his forearm crushed between two strings of drill pipe. A few weeks ago, Hippy nearly lost the toes on one foot when he was tripping pipe and a string of drill pipe dropped wide of the mouse hole on the drill floor and landed on the tip of his boot. He should have lost the toes. His steel-toed boot saved them.

There are a number of reasons work on the rig is as dangerous as it is. First, about everything is made of heavy-guage steel or iron and weighs twice as much as it appears. In a short time most of the men know by heart how much things weigh—drill pipe equals five hundred pounds, a stick of eleven-inch casing equals fifteen hundred pounds, the risers are seventeen hundred pounds, and the styrofoam flotation sheets around each riser are seven hundred pounds. The rig weighs ten thousand tons; each anchor, fifteen tons. Each link on the anchor chain weighs one hundred pounds and is about a foot and a half long. There are eight anchors set with a mile of chain for each. This is one of the favorite facts.

I am impressed by the numbers, too, but I have my own set of guages. For instance, I've never heard anything clang when it hits the deck or strikes another object. It clunks. It rarely bounces. It hits and stays. I can barely pick up a twenty-five-ton tested shackle. Two men at least are needed to move the fifty-ton tested shackles and three are required for the seventy-five-tonners.

It is easy to be carried away by numbers in the oil field. They come in grosses, some so large they lose meaning. Plaisted Oil Company spent $200 million for the right to explore one block in the Mobile South Area 2, and abandoned it after drilling one dry hole. *Gulf Star 45*, our own rig, cost $45 million to build. Before we are done, the Meta Petroleum Company will spend $2.6 million to drill this one hole. Meta leased Block 6—

for $6.6 million. These figures we idly toss around to kill time.

But weights are important if only as reminders of our own puniness, of what could happen if, for a second, we did not pay attention. For we are constantly dealing with gross weights. Because everything weighs so much, and because a large part of our job is moving objects around the rig, we depend on the crane to lift them. And once an object is in the air, if only a few inches, it can drop on you. Thus, General Safety Rule No. 13: *Do not walk or stand under hanging loads.*

Which leads to the next hazzard—the rig as a "floater." Unlike the more conventional "jackup" rigs which have their legs fixed to the bottom, the semisubmersible floats on enormous pontoons submerged thirty feet below the water line. Ostensibly thirty feet puts them below the turbulence and work of the seas, giving stability to the massive rig above, and for the most part this is true. In normal seas the rig is fairly steady, but even then there is a perceptible heaving, and there are times when the seas are too rough to be offset by reballasting and the drilling has to be shut down. It is the penalty that is paid for working in hard waters such as the North Sea or the Atlantic Ocean. At an average $45,000 a day to maintain the rig, shutting down can be expensive. But to drill ahead on a severely rocking rig can also be costly, not only in damaged equipment but in lives. Shutting down due to weather in the Gulf is rare, except in the late summer hurricane weather.

But even the slight heaving can be dangerous—because it is so slight. It is easy to forget it is happening, just as it is easy to forget that an object like a pallet of chemicals, once in the air, acts like a drop line. It wants to come straight down regardless of where the deck is—or your foot may be. And with its weight it is difficult to stop instantly.

Which raises the third, and perhaps most imposing, liability —the other guy. There is no general safety rule telling us to watch out for what the other person is doing. There doesn't have to be one. It is drummed into your head from the beginning. You never trust the other guy—not implicitly. I know that

Pop is careful, that he is not going to screw up through laziness, nonchalance, or overconfidence. Still, he could make a mistake; he could think I was ready when I wasn't and signal Wade to lift while I had my hand inside the chain, which is now stretched over the pallet of chemicals. And yet I have to trust that Pop has handled his end right, that he's got the bar hooked under the pallet, that it won't slide sideways and tip the load of sacks over on me. And he has to have the same worries about me, or he'd better have. I still make mistakes. Then there's Wade. He's good on the crane, deft, nerveless. I trust him. I know he's been in tight situations and hasn't panicked. He's so good I could worry about that. He's liable to act on experience and not pay attention to what is happening right now. There are few things worse than a crane operator who thinks he knows more than the men on the deck. He is one hundred feet away from the load; we are right there. We can see things he can't.

One of the things he can't see is me. I am behind one pallet and in front of another, in a space of about two feet. If he hasn't brought the headache ball directly over the center of the pallet, if he hasn't "boomed up" high enough, then as he lifts, the pallet will swing in abruptly on me and drive my legs into the pallet behind me. General Safety Rule No. 14, *Keep your hands and feet from between two objects,* is not easily obeyed.

"You all set back there?" Pop yells at me. Because Pop and Wade can see each other Pop is the flagman. It is a job he enjoys and, as is his way, one to which he brings a zealous importance. Watching Pop flag a crane operator is like watching a rookie umpire. All his signals are exaggerated and abundant.

The signals are few and descriptive: to lower the crane's boom (and move the hook farther out)—thumb down; to raise the boom (and bring the hook back toward the crane)—thumb up; to move the boom horizontally—stretch the arm in the desired direction and point with the index finger; to lift the load —raise an arm directly overhead, again index finger extended, and swing the arm in a tight circle; to lower the load—stick an arm to the side, cock the forearm down, and point at the deck, again circling. Those are the basic signals. However, there are signals for degrees which are equally essential. If you want the load to go up or down slowly, as Pop will with me standing where I am, he will first point up, then will rub his thumb, middle, and index fingers together in the standard "that's expensive" gesture. If he wants an even slower lift, he will hold both hands up, palms together, and will rub them together supplicantly. The same signals are used for setting a load down.

"Take'er up, Pop," I yell. I plant my feet against the pallet to my back, my shoulder against the bags in front of me. I hold the chains apart as far as I can, waiting for them to draw up tight. I listen for the roar of the diesel. I look up at the ball overhead.

"Boom him up, Pop" I yell. The ball is over my head, not over the pallet.

Pop clenches his upraised fist to stop Wade, gives the thumb up signal. The boom rises; the ball moves in.

"That's good," I yell. Pop rubs his palms together. I feel the tension on the chain. I brace myself. The pallet lifts, wants to swing in. I push back.

"Get it out of here," I yell. Pop begins to circle his upraised finger frantically. The diesel roars and the pallet rises over my head. I relax and move out from underneath it.

Now the hard part, getting it through the hatch and into the mud room. Pop gets on one side of the hatch, I on the other. Wade lowers the load until it is just even with the rim.

"Get outa the way down there," Pop yells. Looking back up at us are Clayton, the derrickman, Bear, and Hippy. Hippy is sitting on the fork lift. They are well out of the way.

"Put her down easy," Clayton yells.

"Turn the damned thing around," Hippy yells. "Don't break none of those bags," he adds. Bear just stands there, arms crossed on his Buddha belly, lips closed, cheeks distended, a collection of wet, brown tobacco splats at his feet.

"Look out for the corner," Pop yells. I reach as far out as

I can over the open hatch and try to grab one of the chains to pull it toward me.

"For Christ's sake, don't push it, Pop," I yell. Too late. Pop has shoved his end over the rim. I can't hold on to my side. It pulls out of my grip. I'm about to fall down the hole twenty feet to the floor of the mud room. Thank God the pallet has the weight it does. I can push off against it. Even though it is moving away from me it offers enough resistance to stop my fall and let me grab the rim of the hatch and pull myself back.

"Boom him down," I yell and see that Pop is already signalling Wade to take the load back up.

This time Wade lowers it exactly over the hold. Pop and I steady it. Pop signals down. The load drops slowly. Too slowly. The rig rocks. One of the bags catches on the hatch and rips open. The dry powder begins to pour out. The draft from below and the wind on the deck picks up the spilling powder and blows it in my face. The chemical is Q-Broxin. It is awful stuff. Unlike the caustic soda which will burn holes in the skin, it isn't harmful—just messy. The minute it gets wet it turns into a thick, sticky paste. I am wet; my clothes are wet. I can feel my face turning brown.

"Idiots," yells Hippy.

"Your ass," I yell back at him. He looks up and sees me and starts to laugh. There's even a smile on Bear's face. I want to get angry at someone but unfortunately it is a no-fault situation, so I curse Q-Broxin for being what it is, wipe the paste out of my eyes, and follow Pop back to the next pallet.

"Watch out for this one," Pop says. With reason. It's a double load and the top pallet looks like a downhill racer about ready to start a run.

"We going to be able to get them together without dropping it?" I ask Pop. Pop isn't sure and turns to Wade. Wade holds up two fingers. He wants them both.

The decision made, Pop now begins to doubt its wisdom. Wade has removed the responsibility from us. I tend to agree with Pop. I think we'll lose the top pallet, too, but I can see

Wade's reasoning. It would be nearly impossible to get the bars under that top pallet the way it is leaning. Obviously, the two were not brought up together. The second was set on the first later and haphazardly at that.

We slip the bars under the bottom pallet, carefully line the chains up. The top pallet is cockeyed; one of the chains won't hold on the edge of the bags; the pressure seems sure to shove the bags askew. We can only hope the other chains will compact the bags enough to hold them all together.

"Take it up, Pop," I yell.

"Get out of the way," Pop yells back, needlessly.

"Look out for that corner!" I yell.

The whole load has swung over and caught on another pallet. It is tipping precariously. Pop frantically signals down. Wade lowers the load but the corner is firmly planted in a bag. It stays. The load falls over the other way. Wade tries to snap it back up, but the chains have slipped. They've moved around the edges of the top pallet. The downward slide begins, bags tumbling one after the other in an avalanche of Q-Broxin to the deck, splitting and pouring mounds of the stuff everywhere.

"I knew it wouldn't work," says Pop, helpfully.

"Guess what we'll be doing after we eat?" I add.

The horn blasts from the crane. Wade signals to take the remaining pallet up. With bags still falling and Q-Broxin still pouring, he lifts the pallet clear and swings it to an open hatch. Very few things faze Wade, especially when he'd rather be out of the weather and back in the day office reading his book.

"When y'all get through eating, clean up them split bags there," Wade announced as he gets up from the table. "Scrub it up real good before it sticks to the deck."

I'm inclined to have another cup of coffee. It is now midnight. We have five and a half hours to go before the tour is over and we can go to bed at last. On the other hand, Pop is feeling uncomfortable. The minute the boss gets up, Pop thinks he should, too.

"Have another cup of coffee," I order him, pointing out that we have only been down here for ten minutes, that Wade is merely going to the day office to sit and won't be seen for most of the remainder of the night, and that our leaving before the roughnecks is a poor example to set.

"Well, I ought to check on my clothes," Pop offers. Pop had finally gotten them in the wash just before eating, but he won't relax until the piles of clothes are stacked once more in front of the proper doors.

"Save it, Pop, for God's sake!" I enjoin him. He fetches another cup of coffee and settles back.

I used to get up from meals early and hustle back on deck only to wait around with nothing to do. I also know that there will be plenty of times in the next six and a half days when we'll be too busy to have breaks; meals will be eaten when we can get them in, and then on the run. It isn't a question of exercising one's right to a half-hour meal break; it's a matter of proportion, of keeping things in balance.

I'm not the only person willing to drag the meal out. Bear can hardly keep his head up. Catfish and Clayton, the derrick-man, are frozen into the oil-field stare. Even Hippy is hiding yawns behind his hand. Only Otis, the mechanic, seems wide awake. To his delight he has uncovered a problem in one of the generators and he is describing in minute detail his diagnosis to Don, the welder, who has lent him half an ear. It's been a long day, close to thirty-six hours in my case. There are still six more hours to go on this tour and another hundred and sixty more or less before we'll see the bank again.

"Well, Pop," I say, "Shall we head up?"

"I'll meet you up there," he says. As if on a signal we've all moved away from the table, dumped our dirty dishes in the tray by the kitchen, and fanned out to our various jobs on the rig.

Retracing my steps down the corridor by ballast control, I notice that the roar of the rig generators and the hum from the thyristor room are not as deafening as they were eight hours ago. And the smell of the waste treatment plant is not as pungent and the odor of diesel fuel has almost disappeared. Oddly, I am not as weary as I was. I'm suddenly feeling at home. The acclimatization is not quite complete, however. The bunk on which my hard hat lies looks inviting, especially in contrast to the pile of Q-Broxin I know is waiting up on deck.

"What do you want to do with this stuff?"

"Can we save any of it?" Pop wonders.

"It's all pretty wet."

"Put it overboard, then. There's nothing else to do." So we do, manhandling oozing bag after bag to the edge of the rig and dropping them seventy feet to the sea below, silently counting the seconds before the thunk of the bag hitting the water returns to us. It's not hard work, just dirty. We can't help but become filthy. We take our time. There's nothing pending. We have five hours. If something should come up to interrupt us, we'd have work to return to.

"Hey, come here." Wade has materialized again. How long he's been standing there I don't know. Pop has disappeared again to the laundry room and in the absence of anyone to talk to I have drifted off into a thoughtless daze, watching the moon trying to break through a clearing sky, as I hose the remaining Q-Broxin down the deck and off the side of the rig. Obediently, I drop the hose and follow him.

"Where's Pop?" Wade asks as he walks away from me.

"Laundry room," I answer.

"Taking a crap?"

"Who knows what he does down there," I say, staring at the ever-present paperback in Wade's right rear pocket.

"Are they drilling or what?" I ask conversationally. I should know, but I don't. And I am only mildly curious.

When I first came on the rig, I brought certain assumptions with me, namely, that on a wildcat rig like '45 finding oil would be the first and only thing on everyone's mind. It took me about two days to discover otherwise. To my surprise, oil—and gas, for

you never know what you will hit when you spud in a well—
was the most underused word in rig language. No one talked
about it. No one seemed to care if we hit any or not.

Once I was in the day office—both toolpushers were there,
one doing some paperwork; the other reading *Looking for Mr.
Goodbar.*

"Grady," I asked, "do you care if we hit oil?"

Their dumbfounded silence was answer enough. I felt suddenly as though I had asked a couple of prostitutes whether
they enjoyed sex.

"Let's hear it for the 'earl,' " Dan Walton, the night toolpusher, had laughed, looking up from his sheets.

"If there's one thing I never hope to see it's oil," Grady had
said, setting his book down. "If I ever see it, you'll see this old
man first over the edge. I won't be staying around."

". . . and I'm going to be close after you," Dan added. To
both men, oil meant danger, a blowout, fire, death. Only tan-
gentially did it mean livelihood.

"We get paid to drill a hole. It don't matter none to us if
it's dry or not. We get paid the same," Grady had explained. "I
guess he might care some though," Grady added nodding at
Elmo, the company man, whom I had not noticed sitting in the
far corner, tipped back in his chair staring up at the derrick.

"Ooo, somewhat," Elmo had admitted, "but mostly I care
you boys don't fuck up and get us killed."

If even the company man, the oil company's representative
on the rig, the man in charge of making certain that oil or gas
is found (if there is any around) didn't care, it was small wonder
that none of us on deck gave it much thought.

Yet, in that room there were three men with an aggregate
total of sixty years experience in the oil fields, whose entire lives
had been given over to the pursuit of oil, who never had and
never would do anything else—nor would they have wanted or
been able to imagine doing any other work.

"Drilling?—not by a half," Wade says. "They got a crater
blowed out down there and they're tryin' to pack it with a pill."

That explains why we haven't been putting drill pipe on the catwalk. It also explains why Grady and Clyde, the driller, were so edgy at lunch. This could be dangerous. We've lost circulation. While drilling, they've cut through a weak spot, a fault, more than likely porous sand, and the mud is escaping into it and not flowing back up to the rig. There is the possibility, slim but present, of a blowout. Not surprisingly, no one is talking about it.

The "mud," a mixture of water and pulverized barite, is the cornerstone of drilling. It does three things: it lubricates the bit, keeping it cool; it washes the cuttings out of the hole; and, most importantly, it stabilizes the upward flow of any gas or oil that might be drilled through. It is the first line of defense. Through the addition of various chemicals, such as the Q-Broxin we have been dumping overboard, the mud's weight and viscosity can be adjusted to accomodate what's being drilled through. A drop in mud pressure is a warning signal that is immediately heeded.

So, all drilling has stopped. A "pill" of fibrous materials such as granulated nut shells, shredded cellophane, or cotton seed hulls is being made up in the mud room and pumped down the drill pipe into the hold. The idea is to pack the crater so the drilling mud will again pass up by it. All this is taking place nearly a mile below us.

"They ever going to get this hole drilled?" I ask Wade.

"They ain't doin' so good, I guarantee you," Wade says. "Company man told me it's costing Meta $240 a foot so far, says it usually costs 'em $60. Don't bother me none. Y'all can stay here forever. I got me twelve fuckin' hours and I'm gone."

The way things are going, we may be here a long time. Last hitch we had returned from our week off to find everything out of the hole, and the rig shut down tight. When we'd left, the drilling ahead was almost on schedule. We had been at just over 2,000 feet. The contract depth called for 11,500 feet. There was a chance the hole might be finished by the time we got back. Rumor had it we'd be taking the rig into Port Arthur, Texas, to "stack up" in dry dock for repairs, and the men were beginning to talk about "going across the creek" to Brazil. . . .

And now, there were all the risers stacked on deck; the blowout preventer, or BOP stack, was stowed near the sub sea room next to the "moon pool," and the deck was littered with parts.

The rig was almost frantic with activity. Being shut down is an oil company's nightmare. The drilling may be stopped, but the money goes out just as fast, twice as fast, in fact, because for every day of no drilling an extra day is needed to make up for it, and both days cost Meta $45,000 each.

So the pressure had built, first at Meta's main office in Amarillo, Texas, then at its offshore office in Houston, from which it doubled on its way to the company man on the rig, who had loaded it on the toolpusher, who had come down hard on the driller, who in turn fired it through the roughnecks down to the roustabouts. Senses of humor were temporarily stowed, heads lowered, asses raised even higher. Breaks became something that used to be—and overtime increased proportionally.

It is bad enough being shut down once, but this was the second time in less than a month. Two weeks earlier, when the rig was being towed onto location, the company man had misjudged the currents and the wind, had ordered the anchors dropped too early, had not sequenced them properly, and the rig had drifted too far off the spot designated by Meta's geologists for spudding in the hole. You do not just reel in eight miles of chain and eight thirty-thousand-pound anchors; it takes special boats to move them. Two days were wasted bringing the rig back.

Why was that the company man's error and not the toolpusher's? After all, who is in charge of the rig if it's not the toolpusher? The answer is simple but delicate. A drilling rig has only one purpose—to drill. As the representative of the drilling company it is the toolpusher's job to maintain the rig so that it can drill. This is his only job, "keeping up the iron." Everything else is the oil company's responsibility, even down to how fast or slow the drilling is done. Many company men are former toolpushers and will listen to the toolpusher's advice. But with the increased technology of drilling, there is a growing number of "educated" company men whose rig knowledge is limited. The wiser of these still listen to the toolpusher, but there are those who presume on their position and view toolpushers as so much "oil field trash."

"I told 'im," Grady said afterwards. "I told 'im which anchors to set down first. He said it didn't make any difference. What am I going to say? He's the company man. We get paid just the same. It didn't make any difference to me. Just as long as none of my men gets hurt. . . ."

And that's how Grady felt the second time when the same company man decided to connect the BOP stack and all the risers on casing that was sticking too far out of the hole.

"I told 'im it wasn't going to hold. I told 'im it was going to snap. You know what the BOP stack weighs? Two-hundred-thousand pounds. You know what them risers weigh? Over two-hundred-thousand pounds. You know how much movement there is down three hundred and sixty feet? That's why we got that ball joint down there. Only with the casing sticking out so far you can't get the stack locked up tight. So, it's going to move, too. I can see he's going to ruin the joint anyway and have to come out of the hole. At worst he's going to lose the whole thing and we'll have to send divers down and fish it out. I told him to send the divers down and cut off the casing. What's he paying the divers a thousand dollars a day for anyway? He says no, it'll work. He's in a big hurry. So the whole thing cracks the first little storm that comes, the ball joint's ruined, we gotta get a new one out here. You know what it's all gonna cost Meta? I could retire on it."

Meta got a new company man; we got the pipe back in the hole; and now, a week later, this.

"Set up the pump and empty these barrels there," Wade says.

It's time to refuel the generators. There is a line of blue, Chevron, fifty-five-gallon drums on their sides under the catwalk.

"You want them all?" I ask.

"Just keep pumping 'em till I tell you to stop," Wade says. I quickly estimate this could take the rest of the night.

"Get Pop to help you whenever he comes back," Wade adds. "Around three o'clock, go down and get yourselves some coffee." And he vanishes once more.

I'm not sure what help Pop will be once he finally returns from the laundry room. The only strenuous part about pumping diesel fuel is moving the barrels and tipping them upright. Normally, I could do it myself, but there is a layer of drill mud on the deck, and drill mud can be as slippery as ice. It is hard enough keeping yourself upright; upending the damned barrels is akin to Sisyphus' travails.

I want to get angry. One would think that on a $45 million rig, someone would have designed a more sophisticated method of fueling the generators. "So this is the Cadillac of the Gulf rigs," I mumble to myself as another barrel slips out from in front of me sending me spread eagle onto the gum muddy deck. ". . . Pop, Goddamn your clean clothes!" I grumble as another barrel stalls on the tip of a small incline, spins around, and rolls back across the deck. ". . . Freeman, I hope they give you life for whatever it was you had to go to court for," I grunt as I try to loosen a cap with a pipe wrench which is too small. I find salvation in counting the hours before Wade will be gone. Somehow the thought adds just the right degree of strength and patience.

Finally I have eight upright barrels surrounding the intake hole, the caps are off or loosened, the suction pump is sucking away, and here comes old Pop, shuffling toward me.

"Got the clothes done?"

"You'd think with all the money they spend out here they could keep them driers working," Pop grumbles.

"Hey, Pop, it's okay you want to fuck the dog around a bit. I got everything under control up here."

"I got back as fast as I could. Them two port driers, you gotta run 'em twice. . . ." Pop is being defensive. He must be tiring. "Hey, Pop, don't worry . . . I mean Wade was up looking for you three times."

"What'd he want? What'd he say?"

"I don't know. All he wanted to know was where that 'God Dah-yumed old man' was."

"You knew; you could have told 'im."

I can't believe Pop is rising to the bait so readily. Pop, the perfect servant, is sweating.

"I did tell him, Pop, but all he did was smile. All I can say is, those clothes better be awful damned clean."

I can see he is about to launch into a defense of the last half hour. I feel slightly ashamed of myself and hold up my hand. "I'm pulling your chain, Pop. He wants us to pump these barrels. That's all. Why don't you go get a cup of coffee, then I'll go down. I've got it here." Pop hesistates.

"He told us to. At three. It's five to. Go on, for Christ's sake!"

"He's probably in the day office. I'll stop in and tell him. I'll be right back."

"Don't hurry. I'm fine."

The weather is clearing. The light rain has stopped. The wind has died down. There are even a few stars periodically visible in the quilted overcast. And it's actually warming up. I take off my gloves, lay them on one of the barrels, tip back my hard hat to let the air dry the hair on my forehead, and take out a cigarette.

There is something pleasant about working at night, particularly toward the early morning. Even with the roar of the rig I have a feeling of repose, as though everyone else is asleep and because I am not, the night and everything about it is mine.

I lean against the nearby rail and stare down through what is called the "moon pool," a large square hole in the middle of the rig through which the drill pipe, encased in broad, round, white risers, drops from the drill floor to the Gulf floor three hundred and sixty feet below. It is a peaceful place to stare into right now: a black square framing a dark blue, working sea and a perfectly centered, straight, white tube held marvelously still despite its extreme length. A geometric feat. A three dimensional Mondrian. Its simplicity stands in marked contrast to the roaring, cold iron monster about it.

I think I hear someone calling my name. I look around and can't find anyone.

"Hey! . . . Perfessor!"

I look up and there on a small deck on the other side of the moon pool, directly under the drill floor are Hippy and Catfish.

"Get y'ass over hear and give us a hand."

Roustabouts do what they are told. I check my suction pump, then walk around to the stairs leading up to the drill floor, duck under a rail to the deck, and join them.

"Put this on, Perfessor," Catfish says and heaves a life vest at me. My feet turn cold. I don't like what I think is about to happen.

"Okay. Let's git this plank here across that opening there," Hippy says. "Keep a good hold on 'er so's she don't go down."

There is a twenty-foot, two-by-twelve-inch, cypress plank resting on the rail. I grab the far end, Catfish and Hippy the middle. We shove and tug it, foot by foot, then inch by inch. It's heavy and gets heavier as we nose it out over the open moon pool, pressing down as we shove to keep the tip end up.

"Heave it now, men," Hippy grunts. The three of us shove as hard as we can. The plank shoots forward and drops down with an inch to spare over the opposite rail. We work it forward until there is about a foot overhang on both sides. "That's good. Leave 'er," Hippy says. "Heave one of them jackets, will ya?"

Hippy slips into the life vest, then reaches down by his feet and picks up a pair of adjustable wrenches, which he gives to me.

"We got to adjust that tensioner there," Hippy explains. "All you got to do is hold these. I'll take care of the rest. You ready?"

No, I state firmly to myself. "Go ahead," I say to Hippy.

"Don't you worry none, Perfessor," Catfish says. "It's no different from walkin' down the sidewalk."

"Your ass, Catfish," I mumble, well aware of where I'm going and where he's staying. Catfish is the same guy who fell off the plank while running risers down in the moon pool. His safety belt kept him from dropping the seventy feet to the water. When they hauled him back up, he took off the belt and the life vest and refused to ever run risers again. He has, but he doesn't like it. That's why I'm here. I'm sure Clyde has had some say in it, too.

"Well, c'mon," says Hippy. The sonuvabitch is dancing in the middle of the board. He has a perfect set of teeth. I can see them all. "Just don't look down."

Don't tell a bear to shit in the woods, I think. How am I not going to look down? And it's ninety feet, all straight. My hands are clammy. I want to put on a safety belt. General Safety Rule Number 20 says I have to, but no one does, at least not on a one-shot job like this. Maybe when you're tripping pipe and working the derrick or the middle board or running risers in the moon pool—jobs where you're leaning out and the belt is the only thing that keeps you from plummeting.

As I timorously shuffle my way out toward Hippy, I vividly recall a conversation I'd had with Grady, the toolpusher.

It was my first hitch on the rig. I didn't know much and understood less. We were on the rig floor looking up at where I am right now—at Hippy hanging these same tensioners.

"What does he have the life jacket on for?" I asked. "If he falls from there, he'll kill himself."

"So we can find the body." Grady answered. "Insurance company won't pay without a body. There was a toolpusher I know who got burned up in a blowout. They never found the body. That was over twenty years ago. His widow hasn't seen a penny yet."

"What do you think you're going to fetch him out with?" I asked, there not being a boat in sight.

"We could throw him a life buoy." Grady answered routinely, knowing well that the chances of a dead Hippy and a life buoy meeting were slim. Anyway, there is only a thirty foot line on the buoy and it is seventy feet from the deck to the water.

"Or we could launch one of the lifeboats," he added as an afterthought.

"Or maybe a passing boat might pick him up sometime," he tried one more time.

"With all the sharks in these waters?" I offered.

Grady shrugged. "Somebody's got to do the job. Hippy likes it."

"Why doesn't he have on a safety belt, at least?" I asked, feeling my questions were falling into the 'Do you care if we hit oil?' category.

"Most of the men don't like them, say they're dangerous; they get in the way," he answered. "Anyway, a man fall from that height, the belt'll probably break his back."

Later I asked Hippy whether it didn't bother him just a little working that high on a rocking rig with nothing to keep him from the long dive. "Motherfuck," he said.

And that's about what I want to say: to Hippy, who is standing in front of me as casually as if he were on a sidewalk, has just taken a cigarette out of his pocket, and has asked me for a light. I've got a wrench in each hand; my body is so tight I'm not sure I can move any part of it without breaking that part off; the plank I'm on is bouncing; the wind is lashing about us; and he wants me to dig into my pocket for a pack of matches. This I don't get paid for. Nowhere in my contract does it say I have to light a cigarette for a Goddamned cowboy on a Goddamned plank ninety Goddamned feet up in the moon pool.

"There," I say, handing him the matches.

"Much obliged, pardner," Hippy says, lights up, and stuffs my matches in his pocket. "Now you just hold this wheel steady and give me them wrenches; we'll git this little job over with. . . ."

"There, that's it. Let's git on back," Hippy says. Slowly I

inch my way back along the plank. There is nothing to hold on
to. "C'mon, babe," Hippy urges. "Ain't nothing to it. Watch."
I know what he's about to do, violate General Safety Rule No.
2: *Practical joking, horseplay, and scuffling is positively prohib-
ited.* I turn, take three long, sure steps, and jump off the end
of the plank down beside Catfish, and turn around. There's
Hippy, jumping up and down as though he were on a spring
board, first on one foot, then on the other, a cigarette dangling
from his mouth.

And there's Catfish laughing. "Wha'd I tell you, Perfessor?
Weren't that bad, unh?" he says.

"See you boys later," I say. "I got some serious work to do,"
and dropping the life vest, slowly descend the stairs and head
back to my pump.

"Go on down 'n have some coffee." Pop is back.

"What time is it?" I ask.

" 'Bout three-thirty. Another couple hours and we can turn
in."

"I don't know, Pop," I say. "The way I feel right now I could
pull a Freeman and fall asleep right here on the barrels."

Pop laughs. "That Freeman! Never seen a man could sleep
like that man, 'cept maybe the fat boy."

"It's hard to say, Pop. Only difference I can see is, Bear
sleeps with his eyes open so you can't see if he's asleep or not;
Freeman shuts everything down. Between pissing, shitting, and
sleeping, I'm surprised he gets anything done."

"Well, we'll work his ass tomorrow," Pop says, "if he ever
shows up."

"He'll show up. He's still got that car to pay off. . . . I guess
I'll get that coffee."

"Two cars," Pop points out. "Take your time."

I shouldn't have gone after that coffee. I don't need it. It's
too late in the tour for a break. As I cross the deck toward the
day office I suddenly feel overpowered with weariness. My legs

are heavy. I have a slight headache. If I go below, I'm not going to want to come back up.

It doesn't help that the vents from the kitchen are along my route. Better than a cup of coffee at this point would be standing by the vents and breathing in an early breakfast. I am reminded of the Scotsman whose girl wanted a box of chocolates, so he took her downtown to the candy store, stood her in the doorway, and told her to breathe deeply.

The last two hours are the hardest to make it through. The last hour is the worst. Twelve hours is a long time. Eight hours are humane, ten hours endurable, but twelve hours . . . especially at night . . . especially at night when there has been nothing to do but make-work . . . I look out at the Sun Oil production rig in the near distance. It is lit up and merry, seemingly bursting with activity. I would rather be there than here. Then I remind myself that that's how we must look to them and the hour is the same for both of us; that probably there are men there looking over this way having the same thoughts; and that dotted throughout the Gulf are brightly lit up rigs manned by weary men quietly checking off the minutes until they are relieved.

The recognition does nothing for my state of mind. The knowledge that other people feel the same way I do, regardless of what that feeling may be, has never impressed me. Because I'm taking a break, I do grab a cup of coffee. I drink it quickly, "borrow" a piece of bacon out of the warming tray, and return above to help Pop watch the diesel fuel run into the tank. I take the long route aft along the port corridor. I notice that Pop's piles of clean clothes are missing. The day shift is dressing. The countdown can begin.

"Pop, what time you have now?"

"Ten after five. The relief ought to be showing up pretty soon."

"Why don't we get these empties out of here."

"Suppose we ought to set up the rest of the full ones. What'd Wade tell you?"

"To keep pumping till he told us to stop."

"That was two hours ago. You see Wade when you was below?"

"He wasn't in the day office. There was no one in the galley. I didn't look in the ballast control. Might as well pump what we've got and let the day shift do the rest."

"Well, we can start up a new barrel for 'em, I suppose."

"Y'all still pumping? They don't want no more." Pop and I look up. There's Wade standing in the darkness in the narrow passage between the diving room and the rail around the top of the moon pool.

"They only needed four, five barrels," Wade says.

"You told us to—" I snap.

"Choke that one right there and put them all back under the catwalk," Wade orders.

"We was about to," Pop says quickly.

Wade turns and walks away.

"Can you imagine what he's going to be like with Brazilians under him?" I mumble to Pop.

"He's good with the crane, though," Pop answers. No wonder Pop has lived so long without ulcers.

"Come on, Pop. The other crew should be coming up. Let's go meet them."

"Let's go," Pop answers. The alacrity with which he drops the pump and forgets Wade's orders is satisfying. I smile as I watch him shuffle off toward the far stairwell. You're all right, Pop, I say to myself. You're a good hand. I can learn something from you.

III

What a wonderful thing is a warm shower! I stand there and let twelve hours of mud and grease and anger and boredom and weariness wash off me and down the drain. I turn up the hot water until it is steaming, cross my arms over my chest, and let the water beat mercilessly on the nape of my neck and pour down my back. I feel no desire to leave its embrace. For the first

time in ages I am relaxed. The longer I stand there, the more
awake I feel. I'm ready to start the day, or at least bury myself
in a book. It is only through force of will that I shut off the tap
and step out and dry off.

"That's good for the soul, Pop," I say as I enter our quarters.
Pop is seated on the edge of his bunk. He has taken his boots
off and his pants, but that's all. He is calmly smoking a cigarette,
his chin resting in the palms of his hands, elbows on his knees.
It's practically the position I'd left him in.

"You look tired, Pop."

"Huh?" he mumbles. "Yeah, yeah." He looks every one of
his fifty-three years. An old fighter who won't quit.

"Go stand under that shower for a while, Pop. It'll make a
new man of you."

"They don't take roustabouts to Brazil, I don't think," he
says out of the blue.

"That's right," I answer, curious. What's gotten into him?

"Maybe I'll see if they'll let me go crane operator. I used
to work cranes lumbering. Ain't done it in twenty years, but if
they'd let me practice a little, it'd come back."

"You can always ask," I say.

"Yeah, I think I will. After breakfast," he says, snuffing his
cigarette out in the ashtray on the chair beside him. "I'll ask
Grady."

Pop, the crane operator. I don't know about that. I'm dubious. There are reasons.

"See you at breakfast, Pop." And I leave him pulling his
shirt off.

"What'll it be, pardner?"

"Two over lightly."

Jackson is exactly where I last saw him, in front of the stove.
Spread out before him on the grill in varying stages of readiness
are patches of scrambled eggs, omelettes, eggs—one and two—
looking at you or over lightly. On the table behind him are a half
dozen small bowls, a pile of chopped ham, another of cubed

cheese, in effect, an assembly line. He reaches for one of the bowls, holds it over the grill and delicately lets two eggs slide out. Quickly he drops the bowl behind him, snatches a plate off the top of a nearby pile while picking up a long spatula.

"Okay, Catfish. One ham and cheese omelette," Jackson yells while deftly flipping the omelette over and over and then onto the plate.

With three long steps he moves to the counter, slides the plate forward and is back before the grill, rolling eggs over, chopping up scrambling eggs, cracking more eggs into small bowls, picking up plates, calling out names, and sliding plates onto the counter.

"There you go, pardner."

I take up my plate, fork out five or six pieces of bacon, some sausage, and a large spoonful of hash brown potatoes. I pass the grits by, knowing I'll hear about it.

There's a place between Clayton, the derrickman, and Dave, the electrician. I set my plate down, go over to the ice chest for a glass, fetch a cup of light coffee, and settle in. This is eating. A man can get fat on breakfasts like these. I pour a glassful of grapefruit juice, shake some catsup over the hash browns, bust the eggs, and let the yolk run in with the catsup in the hashbrowns.

And wait. There is no sense starting in until Clyde has his say. He is sitting diagonally across from me. Without looking up I know his eyes are falling first on the plate, then on me. So, I look up at him, quizzically.

"You ain't never going to learn, are ya, Babe?" Clyde says. "Damn Yankees just don't have no taste." This is all because of the grits, or rather their absence.

"You can't teach them Yankees nothing, can you?" Now Grady has put in his oar. I just have to wait for Carney Ewing, the barge engineer.

"Shit," says Carney.

"That's the word," I reply. "Exactly."

"Yankees!" says Clyde in disgust and goes back to his breakfast. I once tried grits when I first arrived on the rig. No more. There are times when it is good to be a Yankee, if not being able to stomach grits is what being a Yankee means.

"When are we going to start drilling again?" I ask Clayton. As derrickman he is in charge of the mud room. "You got that hole plugged yet?"

"Gettin' there," Clayton answers through, I notice, a mouthful of hashbrowns. "It's comin' back better now. We'll be drilling tonight, I'd guess."

"We're not drilling?" asks Dave, the electrician.

"Nope," says Clayton.

"I'll be damned," Dave says. "Why's that?"

Clayton's mouth is full. He shakes his hand around instead. "Problems," he finally gets out.

"Oh!" Dave says, accepting Clayton's one-word explanation without question. Dave has not been in the oil field long. There is a lot about drilling he doesn't know—and doesn't have to. He is a good electrician.

"See you men tomorrow," I say, getting up.

"I'm on my way, too," Clayton says, rising with me. We're not thinking. He's left handed, I'm right handed. We automatically pick up our plates and turn in those directions. Right into each other. Our plates end up on top of each other.

"You want to take 'em over, that's fine with me, Clayton," I say.

"Oh, I thought you were taking them," he answers.

"You take mine, I'll take yours," I suggest.

"I can't touch yours. You didn't have no grits," he answers.

"I'd never touch one that had 'em," I say.

"Well, I guess we'd just better take care of our own," he says. "Go ahead."

"Fucking Yankees!" Clyde grumbles. Clayton and I look at each other, shake our heads, and walk off toward the kitchen.

I think I'm wide awake. I don't feel like sleeping. I think I'll just get into bed, turn on the light, and read for an hour. Pop isn't back yet. I kill the main light. I am not aware of my feet dissolving. My legs feel light. I stifle a yawn and crack the book at the marker. I read one paragraph and the rest begins to blur. The book gains weight. . . .

"You want that light on?"

I shake my eyes open. There's Pop standing at the edge of my bed.

"Thanks, Pop. Maybe I'll go to sleep."

"Grady says I can give it a try," Pop says.

"Give what a try?" I can barely ask.

"The cranes."

"That's good, Pop," I mumble.

"Tomorrow, maybe."

"Great," I say and roll over. Pop, the crane operator. I'm asleep before the reality of it can penetrate.

5

Second Tour

I

There is a handwritten warning posted in the galley and port
washroom. It reads: "If you want to dance and sing, don't do it
in the corridor. Personnel is sleeping."

The warning is superfluous. There is very little inclination
to dance or sing on *Gulf Star 45*. Certainly not both at the same
time. I'm not sure what difference it would make anyway. The
various rig noises would undoubtedly drown out or at least
shove into the background a full chorus line.

As I lie in my bunk in the dark room, still half asleep, half
daydreaming, I listen to the rig with the concentration and
familiarity of a blind man. There is a symphony going on. It has
taken some time, but I can now discern the various parts, and
I am beginning to hear the story:

. . . the high pitched buzz: that's the crane reverberat-
ing through the heavy steel deck. . . . A muted crash seemingly close enough to make your eyes blink: that's a load
being dropped on the deck. What was it that dropped?
Pipe? A drill bit? A box? Oil barrels? . . . A ping and a
scratch: the hook and shackle, freed up, falling and hitting

101

the deck, the slings being dragged a few feet. . . . A light clang: the sling being removed from the headache ball, the ring being dropped. . . . The buzz again: the boom is going up? Down? Can't tell. What does it mean? Are they putting drill pipe on the catwalk? So maybe we are drilling again, or are they unloading a boat? Will they be done before we get up there? How many boats are there? What's on them? . . . Thum-thum-thum: the compressors. For the barite tanks? . . . Rrrrrr: the fan in the room, the one constant theme, circulating hot air one day, cold the next—a whimsical ventilation system.

. . . Chickk: that's Pop lighting his first cigarette of the day. I open my eyes and roll my head toward the opposite wall. Sure enough. There's a bright red dot at the head of Pop's bunk. And a long exhale. And a short cough. And another.

"Morning, Pop," I bid him.

"What time's it?" he asks.

I turn and squint at the small traveling clock I've set on top of my bed light.

"About four o'clock." Four o'clock! We've slept nearly ten hours, almost half the day. Might as well call it the whole day. We went below in the dark, it'll be verging on darkness when we get back up. For the rest of the week the sun, if there is one, will be a second hand report.

"Wonder if Freeman made it out," I muse.

"He's here," Pop states. "He come busting in here about three, four hours ago. Said he wanted the key to my locker so's he could fetch his gear."

"Good Lord!" I say, "Couldn't he see you were asleep?"

Pop laughs. "I pointed that out to him. He says, 'Hunh?' That Goddamned 'Hunh' of his."

"I got money that says he's in bed right now," I say.

"In bed? I'll bet he's even got his work clothes on so's he can sleep till we go on up," Pop says. I'll bet Pop's right.

"New crane operator's on board," Pop adds. Pop doesn't

have to leave his bed to be a fountain of information. "Come out with Freeman."

"So Wade's gone? Didn't even say goodbye, eh?" I'm not surprised and I'm certainly not disappointed. Whoever this new man is, he has to be an improvement on Wade. What will surprise me, however, is that Wade's name will never be mentioned again. Here is someone who had been with the rig, off and on, practically from its launching, two years ago. Not only that, he was pure Mississippi on a rig that is at least fifty percent Mississippi-manned, and yet, from the point the helicopter lifted him off for the "bank," he ceased to exist. Just like the Old Man, the Bou-Ré playing roughneck, who didn't show up this hitch. We are only parts of this great noisy machine. There are spares on shore waiting for our departure. Maybe that's why Pop doesn't remember names. It makes it easier for the new parts to fit in.

"You going to get up?" I ask.

"In a little while," Pop says. "Maybe I'll just read a little. The old lady gave me this book she says I gotta read. She reads all this fancy stuff which she says I gotta. Course, in that wheelchair she can't do much else so she's got a lotta time."

Pop is a solicitous husband. The "old lady" is a new one for him, the two having been married only a couple of years. She is a paraplegic, victim of a Christmas Eve car crash, twenty years ago. Pop knew her then; she was married to his brother.

Pop flicks on his light, lays his cigarette in the ashtray, leans over the edge of his bunk to his open drawer, pushes around some bottles and underwear, and pulls out a paperback. "You ever heard of this?" He holds up *Chariots of the Gods.*

"Heard of it," I admit. "Never read it." Then I add, "But I've heard it's good."

"I don't know," says Pop, hefting it. "I kinda like that Louis L'Amour."

"He's okay, too," I say diplomatically, not wanting to come down on the side of either husband or wife and not sure there is a great difference between the choices.

"Well, she wants me to try it. I guess I will," Pop concludes.

*   *   *

A treat for supper. Spare ribs cooked "down home" and hot. They are a favorite and we don't get them very often. I load up my plate, sit down, sprinkle some Tabasco sauce on them, fully aware that my digestive tract is going to make me pay later for the indulgence, pick up the first, and bury my teeth in it.

I am not alone. The table is full and we are all going at the ribs like starved pigs. It isn't as though we hadn't eaten. True, the last meal was breakfast at 6 A.M. and it is now 4:30 P.M., but breakfast (What was it Hippy had had?—four fried eggs and grits and hash browns and toast, and Bear had looked at him as though Hippy were on a diet?) had been preceded by two large meals and two "Vy-eeny Weener" breaks. It's just that ribs are ribs, and when you've got them, you eat them until there are no more. And Jackson must know this. The warming pan is heaping again. Reliance Marine Catering must know it, too, which explains why we don't get ribs very often.

"God dah-yum, look at that!"

So I look up and everyone is looking at me. I look around to see whether I am committing some unforgiveable breach of southern etiquette.

"Look at the Yankee eating ribs. Maybe you gon' be all right after all." Who else but Clyde?

"That's funny, Clyde," I say. "I was just thinking maybe you southern boys had finally learned something from us Yankees."

"Hell, you say!" Clyde exclaims. "Ribs is southern."

"Well, that may be, and maybe it isn't," I concede, "but if it is, what I want to know is how you can make them so good and your grits so bad?"

"No, I can see, there's no helping you Yankees," Clyde concludes.

"I can guarantee you, Clyde, we aren't asking," I rejoin, knowing instantly I'm in trouble.

" 'Cept maybe in the winter when it gets co-o-ld?" Hippy tosses in from the far end of the table.

"Someday we're going to let you freeze in the dark," adds Grady.

"We gonna cut you fuckers off," offers Catfish.

This is a losing argument. I've been through it before. Maybe someone else can handle it. I can't. I hold up my greasy hands.

"All right, all right," I get out. "So I'll try grits again."

"Yankees," mumbles Clyde, but he's grinning. So are Grady, Hippy, and Catfish.

* * *

"Ol' Pop. Ol' Pop. How ya doin', Ol' Pop?" Freeman bursts through the door of our room.

"Ol' Pop your ass," Pop grunts, barely looking up from the boot lace he is tying.

"Hunh, Ol' Pop?" says Freeman and gives Pop a gentle punch on his right shoulder, then eases himself down beside him on the bed.

"How you doing, Freeman?" I ask from the edge of my bed. I'm glad to see him back. From the tone of his grunt I suspect Pop is, too.

"Good, man, I'm doin' good," Freeman answers, leisurely leaning back onto Pop's bed, propping his head on the side wall.

"What're you doin' now, Freeman? You fixin' to go to sleep again?" Pop grumbles. "Get your boots offa my bed."

"Ol' Pop . . . you be nice now, ol' man," Freeman says, a big grin on his face. "Ooo, man, I'm tarred. I could sleep the whole night. Hope we ain't got much to do tonight. I ain't had no sleep a'tall."

Pop and I glance at each other, wink, and go back to tying our boots, grinning. We are both saying to ourselves, "Goddamned Freeman. It's going to be one of those nights." His constant weariness is funny at this time of day. We tend to lose our senses of humor toward the end of the tour.

"I can see they didn't put you in jail, Freeman," I say.

"Weren't no problem there," he says, the grin suddenly disappearing.

"What happened anyway?" I ask.

"The judge didn't make no decision. I gotta go back later, that's all."

"What did you do?" I ask.

"Wasn't nothing" is all Freeman will say. I drop it. If anyone can find out, it'll be Pop, and it is just the kind of thing Pop will want to dig out.

"What's the new crane operator like?" I ask.

"Lew? He okay," Freeman answers helpfully. Freeman is not being noncommittal. He probably hasn't the slightest idea, even though he is bunking with him and rode out on the helicopter with him. If Freeman has been true to form, he undoubtedly slept in the trailer until awakened for the chopper, then slept on the ride out, and went right to sleep again on arrival. Freeman is one of the few men on the rig who can sleep the entire time he is off-duty in addition to the snatches he can eke out while on duty. I once asked him what he did when he was home. "I go see my girl, sometimes I work on my car some, but mostly I sleep."

"I guess we'd better get up there," I say. "It's twenty after."

"I hope you're ready to go double-time tonight, Freeman," says Pop, standing up, tucking his can of Copenhagen into his shirt pocket.

"Hunh?" Freeman grunts, pushing himself up out of the bunk.

"Work, Freeman. You ready to work, tonight?" Pop says.

"Ol' Pop." He thinks Pop is funny.

"I'll 'Ol' Pop' you on the side of your hard hat," Pop says.

We head out the door, Pop, me, then Freeman. The second tour is about to begin. What a difference a night's (day's?) sleep makes. I have no idea what is awaiting us, but I'm ready for it.

"This place smell like a shee-it house," Freeman says.

"Hunh?" Pop and I respond in chorus. I don't smell any-thing anymore. Pop can't smell anything anyway. Freeman hasn't become acclimated yet.

*   *   *

"I see you gotta new boss," says Dwayne, one of the Missis-sippi kids. "No more Weed. Y'all real disappointed?"

"I can't tell you how much," I answer. "What's been hap-pening?"

"Y'all got a supply boat due in about twenty minutes."

"You sure you don't want to stay around and help Freeman there?" Pop asks. "He's going to have to go down and hitch up all by himself."

"Hunh?" Freeman blurts. "I ain't goin' down by m'self. No suh."

"Well, we sure aren't stayin' to help none," says Byron, the other half of the Mississippi kids. "Have a good night." And the two of them, followed by Silent Nick disappear down the stair-well, pausing just long enough to let Big Daddy in ahead of them.

"Outa my way, Motherfuckers," says Big Daddy. "I'm hun-gry." Big Daddy is tearing like an eighteen-wheeler headed home. He is also covered with mud, head to toe, as are the other three roughnecks pulled along in his wake. Only Stokes, their driller, is still clean. It is obvious from the look of them that we are drilling again.

"Kinda a little fellow, isn't he?" says Pop.

"For a gorilla, maybe," I answer, then notice that Pop isn't talking about Big Daddy, he's watching Jake, the crane opera-tor, and the new man, who are huddling by the port crane.

"It's hard to say," I offer. In the twilight, appearances are distorted. The new man is shorter than Jake, who is about six feet, but seems taller than Wade was. Slight is the word I might use. He has on cowboy boots, which is unusual on deck, blue jeans, and a light blue cowboy shirt. He is also wearing a new white hard hat like those sold by Gulf Star on the rig. It is tipped back jauntily on his head, and, as he talks with Jake, he makes passes at a lock of loose hair which droops down over his forehead.

Pop and I are both sizing him up. Pop sounds as though he has made some sort of judgment. I don't recall him ever describing Wade as "little." For myself, I am dubious about the cowboy boots and the new hard hat. Why would an old hand need a new hard hat?

A lot depends on what kind of a man this guy is. Our lives are going to be in his hands much of the time. Is he a know-it-all? Does he panic? Is he fast or slow? Bold or cautious? Wade was hard to deal with, but, when he was on the crane, you could feel confident that *he* wasn't going to kill you. If something happened, it was going to be either equipment failure or an unavoidable human error. It would also be nice if we could all get along. Disharmony can stretch twelve hours out unbearably.

Jake and the new man shake hands. As Jake turns toward us I think I distinctly hear him say, "Good luck." He couldn't mean "with this crew," could he? What could Jake have been saying about us? All he knows he would have heard from Wade. I have to admit, we are an odd trio—one old logger and one aging Yankee, neither with much oil field experience, and one Mississippian whose propensity for voiding himself and dozing on a broom is already part of the rig lore. Still, we are becoming a team. We get the job done. Most importantly, we like one another. I think that, starting fresh like this, we can break him in. At any rate, he can't be worse than Wade.

We are about to find out. He follows behind Jake, who passes us by with a pleasant, "Have a good evening" and continues on to the stairwell.

"Well, we best all get started," he says. "Name's Lew." Pop and I both shake his hand and introduce ourselves. "We know each other," he says, smiling at Freeman.

"We got a boat comin' in real soon now. We gotta unload

that and we gotta get some drill pipe up on the catwalk for the drill floor. Why don't we go get that done, then we'll be ready to tie up that boat there.

"I don't know what all else's got to be done tonight, but unless you boys want to I don't see no need in busting our tails," he says as we make our way along the deck toward the starboard crane. "It'll take me a little time to get used to these here cranes, so we'll take it easy, set that pipe up there real slow. Y'all know what we're doing now, don't you?"

What's with this guy? Doesn't he know you don't talk to roustabouts like that? Next thing you know he'll be telling jokes and we'll all be having a good time. What ever happened to the old "heads down, asses up" attitude toward roustabouts? If I'm not mistaken, Pop doesn't seem to be shuffling.

"Pop, you and the Yankee here stay down and hitch up. Freeman, you go up on the catwalk and unhitch. Stand where you're sure I can see you. I don't want no one getting hurt."

*   *   *

The drill pipe is laid out in rows on the forward deck to the starboard of the catwalk. If they were laid end to end they would extend over two miles. Lying side by side, stacked in tiers, they take up a space about fifty by thirty by four feet, each tier separated by odd, broken lengths of two-by-fours.

There are only two things to remember about drill pipe: each stick weighs about five hundred pounds and is round and will roll. When it rolls, it won't stop until it hits another five-hundred-pound pipe, and, if your leg or arm is caught in between, it will be crushed. These are things you learn early and reduce quickly to second nature, for you spend a lot of time around the pipe, not only hitching it up for the lift to the catwalk but using it as extra floor space when unloading cargo from supply boats. When Bear dropped the pipe off the catwalk and nearly hit Pop and me, I was standing on the drill pipe. As I ran away from the falling pipe, I recall being almost more worried about making a false step and breaking an ankle than about being crushed from overhead.

And the rig is rocking tonight. The rain is gone as are the clouds. In every way other than the wind and the seas it is a pleasant night. But the rocking rig has put Pop into an agitated frame of mind, because in order to get the two slings under the six sticks of pipe we want to lift, we are going to have to roll the pipe free. Two men can do this, each at an end shoving. Once moving, the pipe rolls on its own. It also rolls back if left unattended. We have done this many times. Roll one out, then the next, the next . . . until there are six free. Reach under, pull the sling with the hook end through, and catch it back on itself. Once the hook is caught, there should be no problems. The pipe will all cradle together when lifted.

"You ready?" Pop yells. I nod and push. That's one. I go back for the next. It has already rolled a bit free on its own. I push my end. There is a shout from Pop.

"Stop pushing, Goddamn it . . . Goddamn it . . . GODDAMN IT!"

I look up and see Pop jumping up and down like an angry Costello. I suddenly feel the perfect Abbott. I have broken a cardinal rule, I hadn't watched the other man. In pushing my end I had swung Pop's end in and nearly jammed his fingers. I'm going to have to pay the price. Something like this can ruin Pop's day.

"Why don't you pay Goddamned attention? How many times have I got to tell you, check with the other man. Goddamn it!" Pop is really excited.

"Sorry, Pop," I say, sincerely apologetic. "You hurt?"

"No, I'm not hurt, but I coulda gotten hurt. You gotta watch out, Goddamn it."

I stand there, waiting for Pop to calm down. I keep forgetting how nervous he is. Watching him fussing around at the other end of the pipes, I am reminded of one of the unwritten laws of dangerous work: if no one is hurt, then you have to find the joke. How many times have I seen someone brushed with

death immediately become the brunt of everyone else's sense
of humor? That's why Sylvester the Cat and his frequent falls
from high windows provoke such laughter. Pop should know
this. He has worked around danger all his life.

There's a honk from the crane's horn. Lew is gesticulating
behind the cab window.

"Let's go, Pop," I yell.

"Just pay attention, Goddamn it," Pop yells back and gives
his end a shove. The pipe rolls up against the first. We get the
remaining four without incident, hook up, stand back, and signal Lew to take them away.

Wade was a snatcher on the crane. Lew apparently is not.
Wade would get the load off the deck in a hurry, then ease it
over to the catwalk. Lew takes it up slowly, not as high as Wade
would have, then swings it out abruptly.

This takes us by surprise. We have attached tag lines, long
pieces of light rope to the two ends, which we hold to guide the
pipe straight to the catwalk. There is a certain art to handling
tag lines. The six sticks of pipe weigh around three thousand
pounds and once in motion won't be stopped without a tussle.
In the wind they want to turn end to end. We want them to ride
parallel with the catwalk. Lew has moved the load out and away
from us so fast that Pop's tag line has pulled out of his hands.
My reaction is to hold fast. Too fast. I succeed in stopping my
end abruptly, which whips Pop's toward the catwalk, the tag
line trailing after it. I have to make up my mind quickly; either
let go of my line and let the whole load swing 180 degrees so that
Pop and I are trading tag lines, or run down the deck under the
load and try to pull my end back and reverse the swing.

"Let it go," Pop yells. I do. It doesn't work. Pop's line has
caught between two pipes in the rack and the load has stopped
dead and is now pointing at the catwalk. Worse still, because
Lew has not lifted the load high enough, it has drifted into the
catwalk and become wedged.

Pop holds up his arm with a clenched fist. "Hold it there,"
he is signalling Lew. He runs up to free his tag line. I jump up

on the pipe rack and quickly duck foot my way along two pipes to my tag line. As I reach for the end, I hear Pop yell,

"Look out!"

I jump back out of the way of the load which Lew has decided to lower. It misses my head by about three feet.

"What the hell's he doing?" Pop yells and sticks his fist in the air again, jabbing it toward the sky. "Hold it! Hold it!"

We both wait, watching both the load and Lew.

"You got your end?" I yell at Pop. He holds his tag line up. I get directly behind the load with mine and put some tension on it. I raise my left hand, finger up, and swing it. The load rises. I rub my thumb and middle two fingers. It rises slowly. I pull back. The pipe comes free. Pop pulls on his end. The load swings around to parallel, I run across the pipe, my line held slack.

"Don't let it go too far," Pop yells.

I try to pull in on my line and jump off the pipe rack at the same time. There is one pipe in the rack sticking out farther than the others. My shin comes down on it. A streak of pain shoots through my body straight to my hands which are clutching to the tag line. I want to let go and attend to my barked shin. Instead I hold tighter, again stopping my end short and breaking Pop's end loose. This time he grabs it before it can get away.

"Not so Goddamned fast," Pop yells.

"Go . . .," I start, then smother it, aware that my sense of humor is running out as fast as the pain in my leg is swelling. I grit my teeth, haul in on my line. We straighten out the recalcitrant load and Lew lifts it up over the rail of the catwalk. Pop and I let go of the lines. It's all Freeman's now.

But Freeman is having problems of his own, some of which are of his own making. When Pop and I turned the load of pipe over to him, it was in parallel, but as it came over the catwalk it drifted askew, so that my end rode slightly over the far rail. All Freeman has to do is step up and grab the rear end and hold it still and the other end should come back.

But Freeman is wary. If there are two things Freeman

abhors, they are getting dirty and getting hurt. He has on clean gloves and when he grabs that pipe end, his gloves will become greasy. Also, he has paid great attention to the preceding trials, and he is not at all sure what Lew is going to do next. So, he is standing back from the load and is waiting for it to straighten out on its own. Which it refuses to do.

"Grab the tag line, Freeman," Pop shouts.

"Hunh?" yells Freeman.

"The tag line. Grab it."

"Tag line?" Freeman yells. Pop and I both know what he is waiting for.

"I'll go," I yell at Pop and run up the stairs to the front of the catwalk.

By the time I get there, the load of pipe is criss-crossing the catwalk, the end of my line is barely resting on the rail, the other end about to fall back down into Pop's hands.

"Grab your line, Freeman," I shout, snatching at mine. I pull. The load swings in toward both of us. "Signal him," I yell. Freeman points down. Lew drops the load. Too early. My end catches on the rail. The sling goes slack. The pipe begins to tumble down on us. We both leap out of the way.

"Take it up again," I yell. Freeman signals. The sling pulls taut, the load lifts. "Bring him in a little farther." Freeman points at me. The pipes suddenly drive forward like a battering ram.

"Geezuz!" I yell and jump back toward the stairs. The pipes seem to have eyes. They find room between the rail and the edge of the V-door and follow me. I quickly step in behind the V-door feeling like a bullfighter. What is going on here? I look over at Lew. He is smiling, his hands outstretched in a sign of despair. I shake my head and smile back at him. It always seems like a simple job, putting pipe on the catwalk. It rarely is. We'll get it there.

Freeman has a firm hold on his tag line now. He's pulling back. I step out from hiding and grab the end of the pipes and hold on, putting all my weight down on it.

"Lower it, Freeman. We got it there."

Freeman signals. The load drops in a clunk. I unhook my end; Freeman, his. Lew hauls up. The pipe is on the catwalk. We won't have to do that again for a couple of hours.

"Nice job, Perfessor," I hear over my head. There are Catfish, Bear, and Hippy at the top of the V-door.

"Good thing you got Freeman to help you," Bear shouts. If Bear wore sunglasses, he could be a small town sheriff. He doesn't have much use for Freeman.

Lew, the crane operator. Lew, the crane operator. Come in, Lew.

"Where's the bitch box around here?" Lew asks.

"Back in the crane," Pop offers. Lew leaves us to answer the call.

There is a Big Brother aura to the rig. Planted in mysterious places throughout it are intercom speakers. There is practically nowhere you can go and hide from them. Their crackle is as much a sound of the rig as the wheezing of the compressors, the humming of the thyristors and the roar of the generators. They are always filled with an urgency, an immediacy, the portentousness of a ringing telephone in a silent room, particularly since all you usually can hear is the call: *Otis, the mechanic. . . . Dave, the electrician. . . . Clyde, the driller. . . .* Rarely the message: *Lew, the crane operator.* That's the one I worry about because it usually implies: we, the roustabouts. The call looms as a finger of Fate. Apprehensively we wait for Lew to return.

"Let's go tie up that supply boat," Lew says. "It's coming in on the port side."

There's a good sea running. It's hard to say how good from our vantage point, but I can see that the large supply boat is being tossed about as it crosses into view through the swells about a thousand feet off the rig. Pop guesses ten to twelve feet at least. He's probably right. Those boats are two hundred feet long, weigh two hundred tons, and a great deal of that weight is below the water line where the storage tanks for the mud and the water are located. The waves are working her over fairly easily.

"We're going to have to be careful working down there tonight," Pop says.

"This is Freeman's kind of weather, isn't it, Freeman," I say. Freeman is standing beside Pop. He has taken his ever-present toothpick out of his tight beard and is methodically cleaning his teeth.

"I ain't goin' down there nohow." Freeman announces.

"Hell, you ain't," says Pop. "You're going down and we're stayin' here to unload. You gotta have two people here on deck, you know that, Freeman," Pop says.

"Ooo, now, Pop. You know I get sick on those bo'," Freeman says. "What happens I get washed over. You KNOW they never find this boy. Shee-it!"

"Don't you worry, Freeman. You get washed over, we'll write your girl and tell her you went down lookin' good. We'll tell her you were doing the whole job by yourself. She won't ever look at another man again."

"I don' care how many man she look at. She ain't lookin' at me now anyhow," Freeman mutters.

"What? Freeman!" Pop and I say together. I thought there was something strange about Freeman. Now it comes to me. His hair wasn't plaited when he came into our room. Before he comes out for his hitch Freeman always has his hair plaited. This is so that he doesn't have to wash it while he's out here. He just lets the shower drain through the ditches. When you ask him what the first thing he is going to do when he gets home is, he'll tell you: "Wash ma haid." And it is his girl friend who plaits it for him. She is the last person he sees before leaving Prentice, Mississippi.

"She walk on you, Freeman?" I ask.

"She doesn't appreciate those other women of yours?" Pop suggests.

"She's mad 'cause I don't come right over and see her when I get home," Freeman explains.

"Where'd you go instead?" Pop asks.

"Sleep," Freeman answers.

"Here it comes. Let's get ready," Pop says.

There is a struggle going on below us. The supply boat has dropped its bow anchor and is backing in toward the rig. The waves are crashing onto its open stern and running all the way up the deck to the foot of the wheelhouse. It is a battle the boat will win. These boats are capable of towing rigs like *Gulf Star 45* across the Atlantic or as far as New Zealand or the China Sea. They are also used to set the thirty-thousand pound anchors when the "floaters" come on location. Ten-to-twelve foot seas are no more than a nuisance. It's like pitting a heavyweight against a middleweight. Still, it can be a good fight for a while.

"Your haid!" Freeman yells.

I duck while reaching up to ward off and catch the hook from the crane.

Pop has tied a thin rope around the yellow, three-inch thick hawser which is lying in a jumble amidst folds of heavy black hose on the rig deck. The rope has a loop in it which Freeman fits over the hook. Lew starts the cable up. The hawser rises overhead.

"Gi' me a hand here," Pop yells. The hawser rises higher and is stopped by a cord holding it to the rail. Freeman and I hold the hawser still while Pop unties the cord. The hawser itself hangs down toward the water where it is shackled to the middle leg of the rig. Lew swings the boom out, pulling the hawser from our hands. We step back and watch Lew lower it toward the deck of the supply boat. Pop turns on the spotlight by the rail and fixes it on the hawser, picking it out of the darkness for the deck hands.

In lieu of movies or television, the deck hands on the supply

boats provide some of our best entertainment. Barnum and
Bailey clowns could not put on a better show. Nor did the
Christians before the lions which, in a way, they resemble more
closely as we, from our height, leaning on the railing gleefully
looking down, resemble the Romans.

Our urge to hilarity is barely sobered by the recognition of
the danger the deck hands are facing as they wrestle with the
hawser to make it fast to the bollards. Invariably one of us will
comment: "Where do they find these guys?" The answer is, the
same place they find the galley hands, that is, anywhere there
is a willing body.

Lew has laid the hawser across the port gunnel and has
dropped the headache ball and hook directly in the center of
the deck. Using his bow thrusters the captain has backed the
Juanita Candies nearly under the edge of the rig. He is fighting
to hold her there. Black smoke is pouring from the two stacks;
the twin diesels are raging. The turbulence from the seas rolling
between the legs of the rig is fierce. It's a standoff with the
supply boat about to be on the short end if the crew doesn't get
that hawser tied up soon.

Yet the crew is standing back by the winch, poised like a
group of long distance runners waiting for the gun. We can't
hear what the captain is saying to them, but we can see him
standing at the outside controls behind the wheelhouse, and he
is shaking his fist and pointing. Every-so-often one of the crew
will break from the group and dash aft to be met by a breaking
wave from which he beats a quick retreat.

"Go get the damned thing," Pop yells, bouncing up and
down, waving his hands, trying to orchestrate the proceedings
from his position seventy feet above the boat. "What the hell's
wrong with 'em? What do they want?" He flashes the spotlight
on and off and waves it on the deck.

Lew is equally frustrated. He is having his problems keeping the headache ball on the deck. The boat is heaving in the
ten-foot swells, and the ball is bouncing and sliding and gradually moving aft as the captain loses his battle with the seas and

the current, and begins to drift down toward the middle leg. That could be trouble. Lew sits on the horn.

The captain must have made his point. Suddenly the three deck hands race aft and throw themselves on the hawser. They hold on for their lives as a big wave busts over the open stern and washes up the deck, knocking two of the hands off their feet, dragging them forward, then astern like strands of seaweed. They scramble to their feet and work their way toward the bollards. One of them manages a turn on one bollard, slips another over the second. He motions for slack from the other two hands just as another wave hits. The two hands let go of the hawser and leap for the gunnel, nearly falling over it to avoid the sweeping wave. The hawser breaks free of the bollards and starts to drift overboard.

"Don't let it go. Grab it!" Pop yells. He has evidently forgotten they can't hear him. Lew bangs on the horn. Below, the three hands are pointing every which way. The captain is hammering at his horn and waving. The three hands scramble back to the hawser and start tugging at it. I look at Freeman. He is green, I think.

This time they get a bind on the bollards, wrap the hawser three times in a figure eight, and race back to the comparative safety of the winch.

"Let's get the other one," Pop says, and we run forward to the green anchor winch by the day office where the starboard hawser is tied up.

For some reason the deck hands find this hawser easier to handle. Maybe it is because it drops down more directly to the boat and there is less weight to have to haul back, and with only a minimum of confusion the supply boat is at last tied up.

Pop is uneasy, however. "It ain't going to hold. They got it tied up too tight. She's going to snap those lines." Freeman and I shrug. Pop has to have a worry. They are his fodder. "Go wash your clothes, ol' Pop," Freeman says. "We gotta get the hoses on first," Pop answers. He is being very serious tonight.

There are two eight-inch hoses: one for the drill water, the other for the mud. They are a pain in the neck. They are heavy and unwieldy, especially in the wind. Unlike the hawser, most of which hangs down below the rig deck, the hoses lie together in a seeming Gordian knot on the deck by the port burn basket. They have to be raised in two lifts, the last hauling it nearly its length to the tip of the crane's one-hundred-and-twenty foot boom.

We get the water hose over first and down on the boat and then wait while the deck hands search out an extension and make the coupling. While we are waiting to lower the mud hose, Freeman quietly disappears. So does Lew from the crane. We hardly miss them. There is nothing to do for the moment. I'm surprised Pop has not gone down to start his wash, but he is evidently caught up in the odds on the starboard hawser breaking. It has happened before. It would mean we'd have to take a basket ride down to the rig leg and fix a new one. It's a job no one likes because you have to work from the basket which is held precariously close to the water. The waves like to climb the legs and swamp the basket.

The deck hands have coupled the water hose in place and are ready for the mud hose. I look around for Lew and he is back by the foot of the crane talking with Freeman, who apparently is explaining something of great importance. Lew is nodding. Freeman is pointing at his stomach, touching his head, and shifting back and forth. Finally Lew pats Freeman on his shoulder and climbs up into the crane. Freeman ambles over to us, a smile on his face.

"What was that all about?" Pop asks. "He catch you taking a crap?"

"Ooo, no, wasn't that," Freeman answers evasively. Something is up. "We was just talking about unloading that bo'."

"What about unloading that boat, Freeman?" I ask, suddenly putting his act for Lew together.

"Hunh?"

Pop and I glance at each other. The sonuvabitch.

We are right. No sooner is the mud hose hitched up than Lew comes over and announces:

"Old Freeman here says he gets seasick, so you and Pop go

on down on the boat and Freeman here will have to unhitch by himself."

"Seasick?" Pop mumbles.

"And I got these pains right here," Freeman adds as frosting, pointing at his solar plexis. "I gotta see the doct' when I get home." Freeman is keeping a straight face.

"Makes you want to piss a lot, eh, Freeman?" I ask softly.

"Yeh. Do that. I don't feel good at all," he says.

"Maybe Freeman better take the rest of the night off," Pop suggests to Lew.

"What do you say, Freeman? You gonna be able to handle this job?" Lew asks. He is beginning to understand.

"Yeh. I be okay," Freeman says.

"I hope so 'cause you got the next boat, pardner."

No matter how many times I ride it, I quiver before the personnel basket. My legs have learned to be still and not go numb. My palms no longer sweat, and my fingers don't clutch the ropes. My head stays relatively clear and doesn't swim. As I buckle on the life vest and step onto the rim of the basket, take hold of the ropes, and wait for that first jerk to lift us off the landing platform, I am in control of all parts of my body save one—my heart. It won't stop its tattoo.

"You ready?" Pop asks. I nod. "Take 'er up," he yells, jabbing his hand upwards.

"Unh," I grunt reflexively as the ropes come tight and the basket jumps off the platform and swings abruptly out toward the lifeboat. Get the bastard up, I whisper to myself. Another jerk and a third. We are up and over the lifeboat and swinging out ninety feet in the air. I adjust my feet on the rim, just to keep them awake, and lean forward, letting my arms run through the mesh. I'm not quite the leaner-in that Pop is, but close. I am definitely not the leaner-back Hippy and Big Daddy are. I am a confessed horizon watcher. That imprecise measure of distance and depth is comforting. I avoid looking down until I'm halfway there.

Which is soon. For a long time I thought that the crane operator was in absolute control of the speed of the basket, that if he wished he could allow it to free-fall at whatever speed gravity determined. To my relief, I discovered one day that the crane has a governor. Of course, if we should plummet, I'll have no one to blame—if I am lucky enough to survive.

Actually, one of the things you learn riding the basket is, the faster the better, especially when the wind is blowing at you and pushing the basket off plumb, canting it so that your back is slanted toward the water. And there is a stiff wind blowing at us. Thankfully it is blowing against Pop so that when the basket goes off perpendicular, Pop is the one who has to hold on.

This is our first ride with Lew. Wade was fast and smooth. He would get us up and over the lifeboat with room to spare, then sweep us down to the boat where we would hover until we straightened out and he could land us. Jake, the crane operator, is timid. To his credit he remembers when he was a roustabout and how little he liked the basket, only Jake overcompensates. He does everything with a painful slowness. As a result he is a jerker. When Jake's at the controls, you hold on at all times and hope he gets you back on deck before it's time to go to the "bank." Lew starts off like a jerker, but once clear of the rig, he feels more like Wade, and now he has us dangling over the deck of the supply boat as he looks for a safe place to drop us.

The deck is fairly free of cargo. We must be the last stop. Most of the stuff is tied forward meaning that either *Pool 452* or *Ocean Victory* just received a load of casing or drill pipe.

Lew is wary of putting us too close to the stern. He booms out, waits for the boat to rise up, then drops the basket as the boat sinks back. He doesn't want to take the chance of having the boat drop out from under us as we step off. The minute the basket hits the deck, Pop and I leap back, bending our knees to absorb the boat's next heave, and immediately run forward as Lew pops the basket off the deck and swings it aloft.

It is another world down here. We are like strangers. We

even look odd in our heavy boots, bright red life jackets, and hard hats. Smoke from the two stacks is blowing over us, spray from the waves dampens our faces. The rage of the boat's diesels is momentarily deafening but is soon surplanted by the overpowering roar of the rig's generators, the exhausts of which stick out from the rig's underbelly and stream black smoke over our heads. We go forward to greet the inhabitants.

From above they appeared liliputian, but, curiously enough, they are men about our size, and they don't even talk funny. If anything, they are even more foul-mouthed.

"What do you have for us?" Pop asks the mate who is standing just inside the door at the foot of the wheelhouse. Behind him there is a short corridor leading to the galley. An end of the galley table is visible as are two of the crew who appear to be eating supper.

"I guess them barrels is yours . . . and that food locker. We got some milk in the refrigerator here . . . and I don't know what else. We ain't going nowhere else so's I suppose it's all for you."

"What do you want to send up first?" I ask Pop, as we survey the array of oddly shaped pieces of equipment tied down on the deck in front of us: a pair of new bits, a set of bright blue slips, a drill collar, a batch of separators—all drilling equipment. Then there's the large gray food locker and a dozen blue barrels of diesel fuel, nothing dangerous like drill pipe, casing, or pallets of chemicals. We should be out of here in no time, an hour or two at the most.

"Guess he wants the barrels," Pop says. Swinging down from the rig is the heavy cargo net. Pop and I wait back by the galley door until Lew has dropped it on the deck, then we run aft, Pop grabbing the cable, steadying the headache ball which is bouncing over the deck. I unhitch the net and tug on one of the corner loops to straighten the net out on the deck. Pop guides the ball until it is safely over our heads, then he helps with the heavy corded net.

We are working fast now. This is time for hurrying up. It may be my imagination, it could also be the extra concentration, but it seems that the boat is rocking that much more, the waves are running up the stern with greater frequency and size, and the din from the diesels has intensified.

"Hold on," Pop yells. I grab a barrel as a wave washes past us, swirling around and over the tops of my boots.

"Let's go," I yell at Pop. "One. Two. Three. HEAVE." We shove at one of the barrels. It starts to tip over. The boat rolls to port. The barrel wants to twist and follow. "Hold on to it," Pop yells. I brace my foot against a drill collar. All movement is in suspension. The boat rolls back to starboard. "Get it," I yell. We give an extra shove and the barrel tips over onto the cargo net with a dull thunk.

We spin the end around, straighten it out. "Get it into the upper corner," Pop says. I look at him. How the hell am I supposed to do that?

"Get another one quick," I yell. We race back to the next barrel and are soaked by another wave which has flowed in amongst the barrels and risen up as it passed through the tight places. "Goddamn," I grumble. "One. Two. Three. Heave the bastard." It tips over, falls right. We roll it forward and it clunks against the first which has begun to roll back at us. Two down. We get the third . . . the fourth. Pop signals for the hook. "Look out," Pop yells. I instinctively raise up a forearm to ward the ball off. It comes down off the side of my hard hat. I reach for it. Pop is gathering up the net's four loops. I signal Lew to raise the ball up a little and then cradle it between my shoulder and side of my head. We are dancing together on the deck, my boots are filling with water, while Pop snaps the loops over the hook.

"Get it outa here," Pop says and stabs his finger upward. "Watch yourself." I am. I can see what is going to happen. I hold on until there is tension on the cargo net. Lew jerks it off the deck. The four barrels slam together as the net draws around them. I jump back. The net swings in toward me. I climb over the row of barrels just in time. The cargo net slams into the place my knees had just been. "What the hell's he doing?" Pop yells. It's not Lew's fault, it's all part of unloading supply boats.

"Get it outa here!" Pop demands, swinging his upraised hand round and round. I duck. The net swings over my head and up toward the rig floor. Pop and I walk leisurely back to the galley. There are two more loads of barrels to get.

"That was close," Pop says, pinching out a new batch of Copenhagen and tucking it behind his lip.

"Yes, sir," I answer, lighting up a cigarette.

"You want some coffee there's some hot on the stove," one of the deck hands calls to us from the galley. Pop shakes his head, but I believe a cup of coffee is in order. At this rate we are going to miss the nine o'clock break. "There's some cookies up in the cupboard there, brother. Help yourself."

I pour myself some coffee. It is practically coal black, the kind it's said will put hair on your chest. I don't dare ask how long the pot has been simmering on the stove. I doubt whether the Frozen Logger would have drunk this stuff. I try pouring some milk in to dilute it. The milk turns gray. I grab a handful of cookies and go back on deck.

"Here, Pop, eat up," I say and hand him a couple of cookies. He takes them.

"Freeman must be having some trouble," Pop muses. "He is taking a long time. . . . Here it comes. Let's go." I dump the coffee out on the deck, wondering as I run aft how deep a hole it will eat.

The next four barrels are no easier, but we get them. And the next four. It's good to have them out of the way. The barrels are mischievous; they are in collusion with the boat and the seas against us. The remaining cargo should be less obdurate.

"We'll get that food locker next, get that outa the way, then that drill gear," Pop says. "We can take the milk up with us in the basket. . . . Sooner we get offa this, the better. Something's going to break. The captain's not holding her up to the rig. Those lines is the only thing holding her."

"For God's sake, Pop, look at those lines. They tie up ocean-going freighters with lines like those," I say, but Pop may be right. There is a lot of tension on the lines, especially the starboard line, and it's not from the weight of the boat as much as the frequent jerks on them as the boat rises up, then wants to run away from the rig.

All of a sudden our view of the rig is obscured by a white cloud. Almost at once everything is covered with barite dust.

"Outa the way," someone yells, and the three deck hands bolt past us out the galley door, down the deck, and into the dust storm.

"Geezuz!" Pop stammers. "They busted the mud hose, and we ain't got another one."

That's precisely what has happened. Once the deck hands have the valve shut off and the dust has blown away or settled, we can see half the hose dangling from the rig, whipping around in the wind. The other half is still stretched across the deck and into the water.

"They're damned lucky it didn't bust at the boat. I don't know how we'd have gotten it back if that'd happened," Pop says. "Looks like they'll be taking us back up to fix it," he adds with evident relief.

But that's apparently not what's going to happen because here comes the ball with an empty hook. Lew must want us to hitch on this half of the hose. I smile. Ol' Freeman has some work to do—all by himself. Lew's all right. Before this night is through Freeman is going to wish he'd talked less about his stomach.

"Poor Freeman's going to lose some crapping time," Pop says with a certain delight. This felicitous turn of events has temporarily banished his worries about the boat breaking loose. We hitch the hose up and watch it disappear over the edge of the rig, then stand back to watch Freeman begin tugging at the rig's half, hoisting it by hand, foot by foot.

"Come on, Pop. Let's go in and sit down. We've got ourselves a long wait."

The *Juanita Candies'* captain is red-ass. He is pacing back and forth across the wheelhouse. Every French ounce of his Cajun blood is boiling.

"Look at that motherfucking rig, will ya. Fifty million dollar. Fifty million fucking dollar. That there's the biggest and the best rig they got in this Gulf and they ain't got enough hose to tie onto this li'l boat. You think maybe with all that money *Gulf Star* maybe have another hose? No way. You think maybe they can buy fifty more feet? No fucking way. You think maybe . . ."

He is so mad he's bubbling. In the pitch black of the wheel-house he begins searching for a can of soda he's placed some-where. The mate is stretched out on a cot in the far corner. Only the red glow from his cigarette locates him. One of the deck hands is leaning over the chart board. The captain finds his soda on a stool in front of the steering gear. As he reaches for it, the boat lurches. His hand hits the back of the stool spinning it around. The chair bumps into his hip and the soda falls on the floor.

"Sonuvabitch!" he rages, reaching down and retrieving the can. "Somebody clean that up real quick." And he goes to the port door, opens it, and heaves the can out into the water. The deck hand pushes himself off the chart board and heads below for a mop.

"I told them, that's no good. We need more hose. I tell 'em at *Ocean Victory,* they give me more hose. I tell them anything I need, they say, 'Anything you want, Captain, you can have.' Them ODECO rigs, they spend the money, they keep up the equipment. You don't mind working for dem people. *Pool 452,* same way. You want some thing, they say, 'Okay, Captain, you know best.' Dis *Gulf Star.* Cheap. . . . My good God, we never get home now. Five days out in these seas and now this. My Gawd!"

It is probably not the right time to point out that had he used the bow thrusters more diligently and not allowed the boat to drift so far from the rig, the hose might have been long enough and that if he doesn't pay a little more attention, the lines may snap, and if that happens, no matter how long the hoses are, he will be taking them with him. One thing you don't do with a coon-ass captain is argue. And he is probably right, or at least half right about *Gulf Star.*

For the first time all evening there is nothing for us to do, except wait for Freeman, and that could mean we'll be here for hours. There are worse places to be marooned. At least down here there is no one to tell us we can't sit down. As long as one of us keeps watch, the other can sleep. If we had some money, we could play Bou-Ré with the deck hands. There is even plenty of fresh coffee. One of the deck hands has offered his recent copy of *Penthouse*. He also suggested exchanging the boat's library of Louis L'Amour for whatever we have on the rig, a straight switch, sight unseen, the guess being that L'Amour is so prolific the chances of duplication would be slight. He is disappointed when Pop tells him the rig has no such library, that what books exist are privately owned and secreted in various lockers.

Curiously, we can't relax and enjoy this luxury. It really is another world down here. We are rig people, not boat people. It is fun to visit, but not to stay. Pop and I find ourselves pacing, getting up from the galley table, and wandering to the door and looking up; going up into the wheelhouse and standing in the dark gazing at the rig, almost aching to get back on it, to be doing something . . . anything.

What a strange thing to want to be back on! Only the ocean could inspire such a monster. From the rig floor you are not aware of its gargantuan proportions. From the air it forms a tiny square gray patch on the endless water surface. But from below it looms out of the water like a prehistoric spider, its six massive tubular legs thrusting stiffly downward into the Gulf from its thick, gray, horizontal back, odd pieces of esoteric equipment clinging to its sides like barnacles; its towering derrick spine piercing the black sky, at the zenith a red star blinking, a broken shaft of light baring its cold, straight, triangular outline; and its roar, the belching black smoke from its belly, its loud screech as the stabilizers high in the derrick release pressure.

It overwhelms its territory. The muscular little supply boat rolls at its side like a pilot fish, without which the monster would expire. Those black water and mud hoses reaching up to its gut are nothing less than lifelines. How fragile, but how terrifying the rig is! And how utterly impervious it is to the coon-ass captain's bellowing.

I leave him to his harangue. The mate and the deck hands have long since gone below to their rooms. I find Pop propped up on a bench in front of the galley table, thumbing through *Penthouse.* He folds it shut as I enter the galley and places it on the edge of the table.

"It's all right, Pop," I say. "I won't tell anybody."

"What time's it?" he asks.

"I don't know. Midnight? Must be about that. . . . Shove the magazine over here, will you?"

"Are they doing anything?"

"You can't tell. It's hard to see over the railing. I suppose they're doing something. . . . How come all these women all look like my wife? I can never figure that out."

"Oh yeh?" Pop grunts. "Your wife looks like that, what're you doing down here?"

"Gets boring, Pop."

"If I had a wife looked like that, I'd—"

"Basket coming down, men," the captain yells down the gangway.

Pop is off the bench and on his way to the deck almost before the sentence is finished.

"You'd do what, Pop?" I yell after him, the picture of Pop and a *Penthouse* pinup gamboling through my mind. I have another picture of Pop. This one has Pop taking his seventy-five-year-old mother-in-law out for a Sunday drive. Next to Pop is a coffee can for his Copenhagen spit. Next to the can is another, his mother-in-law's. She also chews tobacco. I like this picture better because, according to Pop, that's what happens when he's home.

Pop doesn't answer. His fantasies have been shunted aside by the arrival of the basket. He doesn't bother to buckle his life vest. He cannot contain his eagerness to get off the supply boat and back on the rig. He signals Lew to take it up before I've put both feet on. The sudden jerk knocks my one foot off. I'm hanging off the side as Lew swings the basket up across the gunnels and over the water.

"Mother of God!" I groan, then for the hell of it add as I pull myself erect, "Pay attention to the other guy."

"Hunh?" Pop answers absently. He's staring up at the rig, he hasn't seen anything.

" 'Hunh' yourself, old man," I say. Pop catches himself and smiles.

"Did y'all have a nice vacation down there?" Lew asks. "Poor ol' Freeman here found himself with some work to do." From the slightly incredulous look on his face I guess he is catching on to Freeman.

"We all felt bad just thinking of how Freeman had to work all by hisself," Pop admits.

"Let's go get something to eat," Lew says. "C'mon, Freeman. Put some food in that stomach of yours."

"I could do that," Freeman says, extricating himself from the pile of hose on the deck. His gloves are already black. Even his dark green coveralls are smeared. The toothpick is gone. If I'm not mistaken, his brow is damp. Freeman just may need a shower when the tour is over.

"That captain is awful damned mad down there," I mention as we step over the uncoupled sections of old hose Lew and Freeman have been trying to piece together. "Says he wants to get out of here."

"He knows where he can go," Lew spits. "He can yell all he wants, the crazy coon-ass bastard. If he'd have kept that boat up tight to the rig insteada letting it drift out, that hose woulda never broke. If he doesn't watch it, he's going to bust those lines and he's never going to get to the bank."

"Tell him that," I say.

"You can't tell a coon-ass nothing, pardner."

I file that away. With a name like Millard, Lew has to have some coon-ass blood of his own.

II

Twelve-thirty A.M. The night is passing quickly. The weariness
that clouded last night seems to be gone. I don't recall having
even thought about the time. We missed coffee break and are
an hour late for lunch. (What else do you call the meal that
comes at midnight between supper and breakfast?) Yet, I don't
feel particularly hungry. Which is fortunate because there is not
much left to eat. Clammy spaghetti and lukewarm meatballs. I
put rice on a plate, pour a bowl of gumbo, collect a square of
cold corn bread and take a seat across from Pop and Lew at the
port table. We are the only people in the galley, but Freeman
has sat down at the starboard table.

"Bring your stuff over, Freeman," I suggest as I pass him,
to which he says "Hunh?" and continues eating away the moun-
tain of plain spaghetti he has piled on his plate.

". . . Abu Dhabi, Brazil, you name it, I been there, an' I'll
tell you this, pardner, I'm headed back across the creek; this life
out here ain't no good at all. . . . Hell, I wouldna come out here
but I'm waiting on a job in the North Sea that's comin' up in a
month an' Gulf Star asked me to come out 'til the job's done
'cause then the rig's going to stack up before goin' to Brazil, is
what I hear. . . ."

This is the first time I've had a good look at Lew. Slight is
still the right word, but so is wiry. And leathery. And cocky. The
man has to have some coon-ass blood. He wants to give the
impression that Lew Millard makes his own decisions, that he
would as soon pack his bags as stay around if, for whatever
reason, he did not like being where he was.

"So you're not staying with 45?" Pop asks with uncus-
tomary interest. He has picked up on something.

"I'll tell you something, pardner, from what I've seen of
this rig already I wouldn't want to stay here too long. If ol' C.
W. Chase could see this rig, he'd rise straight outa his grave and
some heads would roll.

"Why's that?" I ask, amazed that so firm an opinion could
have been born from such a short stay.

"Just look at this thing. How old's 45—a couple of years? The damned thing looks ten. This one come outa the yards after 48; they was built together down in Brownsville, and you should see 48. Looks like it did when it come out. There's no reason 45 has to look worn out like this. Why don't they give her a coat of paint?"

"Money," I offer and sit back and wait.

"You gotta be kidding. Money?" Lew bursts. "You talking Chase money, pardner. You know what that is?"

I do, at least as well as most of the men on the rig. However, one of the favorite pastimes along with discoursing on how much things weigh and cost is enumerating the various enterprises collected under the C. W. Chase umbrella. To work for Gulf Star and be able to talk about Chase money is tantamount to being a Chase and having the money. There is a sibling pride mixed with awe in the tone with which the men refer to "the children"—Cassie, Leland, Barton, and Howard. Everyone knows that Barton owns a professional football team, but that he also owns "the largest herd of Charolet cattle in the country" is a pearl to be guarded and saved; it's an ace-in-the hole to be used when someone is coming on too intimately. I can see Lew is champing to reveal his treasures. I'm game. "I've heard stories," I answer.

"I'll tell ya. When they built this rig here, Gulf Star had a contract for five of them with Marathon LeTourneau down in Brownsville, Texas. They was supposed to cost twenty-five million dollars each. Halfway through, Marathon went bankrupt. There was a steel strike or something. The cost went up to fifty million dollars. You know what Gulf Star did? They bought out Marathon. Just bought 'em out. You know how they did that, hoss? They took a loan from their own bank, I mean, one of Chase's banks, and that's how they finished building the rigs."

I didn't know that. Sources closer to the facts claim that Gulf Star bought only forty percent of Marathon LeTourneau, just enough to control the operation. Where Gulf Star got the money is unclear. Lew could be right about that.

"And I'll tell ya more." Lew is warming up. He has a cap-

tive audience. "The Chases own Chase Oil Company and
Plaisted Oil Company. Two separate operations, right? Hell,
that's why down in Houma they're both located in the same
building. Two names on the outside, one floor inside, same
people working for both. And they own Gulf Star Drilling Company. Who does the drilling for Chase and Plaisted Oil? Gulf
Star. And they own Off Shore Marine. Who supplies the boats
for Gulf Star? Off Shore Marine. Who provides the tools? General Rental Tools, owned by the Chases. You know what, pardner? Chase owns Reliance Marine Catering and who does all
the catering for Gulf Star? Reliance Marine. Now you tell me,
do they have money for a coat of paint?"

It turns out that Lew is wrong on at least one point. Unless
there is a "straw" involved, the Chases do not own Reliance
Marine Catering, but they might as well, for Reliance Marine
does have a long term contract with Gulf Star. Barring such
minor factual errors, it does seem Lew has an argument. There
should be enough money for a coat of paint or at least enough
squeegees and scrub brooms to keep the peeling deck clean.

"So you're not going to Brazil?" Pop says. He hasn't been
paying any attention to Lew. He has something on his mind.
Lew shakes his head. "And they take crane operators overseas,
don't they?" Lew nods. I'm beginning to recall something I only
vaguely heard. It's my own voice saying "Pop, the crane operator." I'd forgotten all about Pop's early morning meeting with
Grady.

"I used to run cranes logging back in the state of Washington . . ."

Pop's recollection is cut short by a loud, dull bang and a jolt
which vibrates through the rig.

"What the hell?" Lew says.

*Lew, the crane operator. . . . Lew, the crane operator. . . .
Come in, Lew. . . .*

As Lew comes around the table to the bitch box on the wall
behind me, Pop states:

"That boat broke loose."

Yeh? Lew here.

Yeh, crane operator, that boat there broke loose. Better come up and see what you all can do up here.

Okay, we'll be there. . . . "Well, men, let's go see what happened."

It's just as Pop had predicted. The starboard hawser is in shreds. One short piece still attached to the rig is washing up and down the leg with the swirling waves; the other end is now lying across the deck of the boat which is rolling almost sideways to the waves. The captain is at the outside controls fighting to keep the boat near enough to the rig so that the water hose doesn't pull off.

"Let's get that hose before we lose it, too," Lew yells, mounting to the crane.

For the moment there is nothing for Pop and Freeman and me to do except watch, and be thankful we are not down on the boat. Even seventy feet away may be too close to the coon-ass captain. He is beside himself with anger. He looks like a whirling dervish at the controls, yelling and pointing and wrestling with the gears, trying to get the stern around and back up into the wind.

There is mayhem on deck. The hose is washing back and forth, the headache ball is bouncing. It takes two deck hands to hold it still and even then it pulls out of their hands when the boat plunges. Three times they try to hook the hose and miss. Lew is having his troubles, too. Keeping the ball on the deck of the ornery boat is like threading a needle on a moving train. Once the boat falls downwind as he is dropping the ball and it becomes entangled in the pipes that run along the gunnels. He tries to pull up but the hook is caught and he has to let the cable fall on the deck while the deck hands skittishly work it free.

Almost accidently the hands get the hose hooked and Lew snatches it off the deck and swings it up straight over our heads, then slacks it down on us in such a hurry that both Freeman and I resemble snake dancers with a great black boa draped over our shoulders, down our backs, and around our feet. We don't even attempt to fold it neatly; it is all we can do to keep from being knocked down.

"Get that other line," Lew yells through the crane window.

Again he drops the ball down. The deck hands have already unwrapped the port hawser from the bollards. The hook up is neat and clean and we tie it off on the rail and stand back. The boat moves out from beside the rig and casts off its line from the bouy. She is on her own, but with our food, drill equipment, and half-full tanks of drill water and mud. What started out as a simple two-hour job has now taken nearly six hours and it is only half done.

"They'll wait till morning to get the rest," Pop states with confidence based on common sense and an apparent un-familiarity with the way things are done in the oil field because no sooner has he spoken than Lew motions us over to the crane and announces:

"We're going to try to get the rest of the stuff off now. I want you two to go back on down there. Freeman, you stay here again . . . and watch yourselves down there. Don't hurry any-thing. Wait 'til he gets the boat where you want it. We got all night."

I don't like this at all. Nor does Pop. It's rough down here. Trying to keep our balance on this lurching boat, hitching up with waves bursting over the stern and washing up the deck then sucking our legs back in the undertow, I am dramatically aware of what the deck hands had gone through earlier. From up there we must look like clowns; down here we feel anything but. There is nothing about this that is funny. We want to get out of here. We have to fight back the urge to hurry. It would be easy to make a lazy connection, just to get the piece up and away. Twice I say: "That's good enough," only to have Pop come over and re-do my connection.

"If that lets go, somebody could get hurt," Pop says. He's

right. All I can say is "Sorry." I know better. I tell myself to calm down, take it easy. I am standing there, silently berating myself when a sudden pain runs through my right foot and up my leg.

"Geezuz!" I yell. One of the drill collars has started to roll up my foot. This is why we wear steel-toed boots, only I don't. Every week I'm on shore I promise myself I'll buy some, then I get to the shore and look at the price and tell myself that I live right—it will be the other guy who will need them. Now, I am about to pay the price. The drill collar is mercilessly heavy. It isn't long, about four or five feet, but is made of heavy-guage steel. When drilling, it is hitched to the drill bit on one end and the drill pipe on the other and functions as a plumb bob to keep the bit as nearly perpendicular as possible.

I try to yank my foot away but it is stuck. If the boat goes anymore to starboard I'm done. We sit for what seems like hours on the crest of a wave. The boat could go either way. I am clenching my teeth and helplessly watching the small drama going on at my feet. Then the boat slips off to port and the drill collar rolls back and crunches the side of the boat.

"How'd that get unchained?" I curse, wriggling my toes about in the tip of the boot. They are numb but all right. The pain subsides.

"That leaves the food locker and the milk," Pop says as we watch the drill bit swing up and over the rail of the rig. "Why don't we start getting the milk out on deck."

It's still too early to start relaxing but I sense a yawn waiting in the wings.

"What time do you have, Pop?" I ask.

Pop fishes his watch out of his pants pocket. "Quarter to three."

"Feels like it, doesn't it?" I say, grabbing one of the milk boxes he has hauled from the walk-in cooler and carrying it out on the deck. "How many are there?"

"Looks like fifteen or twenty," he answers. "Watch out for this one; it's leaking."

"Let's get the food locker, Pop."

The food locker stands about eight feet high with eyes on each corner for the shackles. I hoist myself up on it and wait for the double sling to come down to me. It's a game of "dodge 'em" as I try to hold the loose wires from slapping me while I make each corner fast.

"Want some help?" Pop asks solicitously.

"Yeh. Keep the boat from rocking," I answer, tightening up the last shackle. "Look out. I'm coming down. It's all yours."

Pop signals Lew to take it up. Lew misjudges the boat's heave. He gets it up about two feet and begins his swing out. He'd have made it if the boat had gone down as he expected. Instead the boat comes up and the locker slams against the gunnel tipping partially over. It slides up and over and out. What a mess there must be inside! Whatever hasn't been spilled by the rocking boat has to have been dislodged now. It will give us something to do for the rest of the tour once we get back on the rig.

The milk boxes go up by personnel basket. Pop's estimate of twenty is right, which is too many for one load. We send the first up, pile the next bunch on the second, jump on ourselves, and leave the *Juanita Candies* for what Pop and I both hope is the last time tonight. As we ride up to the rig, the yawn arrives and I let it come. I hadn't realized how tired I was.

"That work takes it out of you, don't it?" Pop says.

"You got that right, hoss," I answer, mimicking Bear, who, it occurs to me, I haven't seen tonight. No, that's wrong. I did see him when we were loading drill pipe on the catwalk. We did do that tonight, didn't we? So much has happened time has become blurred. I think I'm ready for bed or at least a cup of coffee. For some reason I feel cold. Of course. I'm soaking wet. Which reminds me.

"Pop, how you planning to get your wash done?"

"All I got to do is dry it. I put it in the wash before we ate."

"You're all right, Pop."

He smiles. He's kind of proud of himself, too.

"You all want to go down and change, go ahead. Then we'll get a cup of coffee. I imagine you boys can figure out how to spend the rest of the night stowing the food there. Maybe you'd better get the milk first, then take your time."

"What're you going to do about that broken line?" Pop asks.

"You want to take another ride down and fix it, Pop?" Lew asks rhetorically. "Maybe ol' Freeman here would like to," he adds. "What do you say, Freeman?"

"Shee-it, man," Freeman groans. "I ain't goin' do that 'lone."

"Ol' Freeman here's had hisself some kind of a night, ain't you, Freeman. Had to put drill pipe on the catwalk all by hisself. I think that stomach of his is going to be real fine tomorrow night. Ain't that right now?" Lew asks.

"I go down that bo', you ask me, man."

"Tomorrow's your night, Freeman," Lew says, winking at Pop and me. "Nah, we'll let the day crew take care of that busted line. The hell with it. We've had enough to do tonight. Let's go get some coffee."

Lew's faith in us is not misplaced. It helps that everything in the food locker is bottom side up and thoroughly shuffled with many of the cartons either so ripped or broken beyond use that we have to pack and unpack and repack to take advantage of what we have. And then with Pop exercising extreme diligence over the drying of his laundry and Freeman making up for lost crapping time and I making sure I get one box below in tact rather than take the chance of losing two, we are indeed capable of stretching out the emptying of the food locker for the remaining two hours of the tour. But just barely. One less crap or a double load below and the day crew might have caught us in the middle of one of Pop's jokes. As it is, they meet us coming down the stairs as they are slowly climbing up, and we can say with a straight face:

"You boys can take it easy, today. We got it all done for you."

"So we heard," Dwayne says. "Except for that busted line."

"Well, that's right. . . . Enjoy."

6

Third Tour

"Yeh. Okay . . . I'll be right there . . . Give me five minutes . . ."

I lie on my bunk, eyes closed, fully wide awake, furious in
a resigned sort of way. What the hell do they want us on deck
for? I wait for Pop to switch on his light, but all I hear is snoring
—loud, raucous snoring that reminds me of a two-man buck
saw. What's happening? I swing my legs over the edge of the
bunk and sit there hunched over. My legs are tired but the rest
of me feels fine and ready for a day's work. I flip on the bed light
and squint at my clock. Eleven o'clock. Damn! Four hours sleep.
Why isn't Pop awake? The way he's snoring he should have half
the timber in the state of Washington bucked up before supper.
Was I the only one called up? I stumble out of the room into my
own pile of dirty clothes. There's no one in the corridor. I open
Lew and Freeman's door a crack. They are both asleep—Freeman dead to the world, an arm and a leg draped over the side
of the bed. I must have been dreaming. If I wasn't, they'll come
and call me again. I tip-toe back to my own bed and crawl in,
flip off the light, and lie there waiting in vain for sleep to return.

139

But how do you sleep in a working lumber mill? If Pop would only turn over on his side or stomach. I can hear him shifting around but he stays on his back. This is awful. I'll go to the bathroom. Maybe by the time I return he'll decide to take a break.

On the way down the corridor I meet Harmon 32, the galley hand with the extra plastic bag on the crew boat ride out. He is wearing one of those three-dollar football jerseys with his name, Harmon (I guess that's his name), stenciled across his shoulders and a big 32 below it. There is nothing of the football player about him. Skinny with a curved spine, an almost skeletal head, his face crossed with tight, gray wrinkles; long, unmuscled, white arms. Age: indeterminate. Definitely Tchoupitoulas Street.

"How you doing?" I ask as I pass by.

He steps aside, his back to the wall, his mop in his hands. He nods servilely.

"What kind of day is it?" I ask.

"Don't know. I ain't been out," he answers. And he's not going to, either, except to dump the trash from the galley into the burn basket. Here he is, buried for the next two weeks in the bowels of *Gulf Star 45,* mopping and waxing floors, scrubbing bathrooms, making our beds. The sun will rise and set fourteen times and if he sees ten total minutes of its passage, he will be lucky. At least for the next two weeks he has a free room and three good meals a day and there is no way he can get a bottle.

"Looks good," I tell him, sweeping my hand down the corridor.

"Yeh, sir," he answers, nodding again.

"We all don't help you much do we?" I say.

"Not much," he agrees, still nodding, and I continue down to the bathroom, do what I have to do, then wander through the locker room to the stairway leading up to the aftermost corner of the rig. While I'm standing there, just looking up at the pale blue sky, there is a monstrous beating of wings, of air being slapped and battered, an all-consuming roar that fills all spaces. It rises to a deafening peak and pulls away. The morning helicopter flight. What new faces will there be at the galley table tonight? I turn around and go back to my room.

Wonder of wonders! Pop has rolled over. He is just wheezing softly now. I crawl under my covers, take up my book . . . and fall back to sleep.

Shrimp tonight. Look at it all. Boiled shrimp in that pan, fried shrimp in the next. Cold shrimp over with the salads. And potato salad. Am I in heaven? I heap my plate and take a seat between Dan, the toolpusher, and Clayton, the derrickman. I thought I was being profligate. I'm a piker compared to these two.

"Hey, Jackson, where are the oysters?" Carney, the barge engineer, hollers through a stuffed mouth, from across the table. Oysters? Where does Carney think he is?

"Ain't got none this time, boss," Jackson yells back from the stove.

"You a lyin' sonuvabitch, Jackson," Carney yells back. I concentrate on my eating. This is getting rough.

"Oh, we got oysters. You right there, Carney, but you don't want to eat 'em," Jackson answers. I look over at him. Jackson is leaning back from the stove, arching his lanky frame so he can see into the galley. He is grinning.

"How you know that for a fact, Jackson?" Carney asks. I now look at Carney. He is stuffing a fried jumbo shrimp into his mouth. He's trying to keep a straight face.

"I tasted one myself and they was bad. Had to throw the whole can out," Jackson explains.

"Hell, you say," Dan, the night toolpusher, chimes in. "You let me check your gear afore you go to the bank. . . . Ol' Jackson there's got a nice meal all tucked away for his kids. Isn't that right, Jackson?"

"Oh, no, sir, that's not right," Jackson says. He is standing at the counter, pouring more shrimp into one of the
warming pans. "Let me explain it this way. . . . The way I
sees it, you gentlemen is mighty lucky you got a cook like
Ol' Jackson here—"

"Your ass," burbles Hippy. Everyone has stopped eating.
They are looking down at their plates, suppressing giggles.

"No, sir, I'm telling you straight an' I'll tell you why . . . 'caus
Ol' Jackson here thinks so much of you gentlemens that he's
willing himself to test the food before he gives it to you. He
don't want none of you getting sick. No, sir. Now I calls that a
real friendly attitude and I think you mighty lucky to have Ol'
Jackson here."

Jackson's dissertation has stopped Grady in the middle of
a bite, half the shrimp inside, the other half outside the mouth.
He doesn't know which way to go, stuff or yank.

"If that's right, how come you wait for the night cook 'fore
you eat, Jackson?" yells Catfish.

"Ain't that right?" Grady shoves around the shrimp. "What
you say to that, Jackson?"

"I say that's another sign of my generosity, letting you
gentlemens have the best," Jackson counters.

"Well, we sure appreciate it, Jackson," Carney says. "We
sure do."

"Just don't let me catch you with that can of oysters, Jackson," Dan, the night toolpusher, says.

"No, sir, Dan, I won't let you do that," Jackson says.

"I'll bet you won't," Clyde, the driller, grumbles.

It's a boy. Dave, the electrician, is lost for the rest of the
night. His first grandchild and a boy grandchild at that. He just
heard about it and the news has left him weak. He is sitting in
the easy chair down at ballast control, where I've been sent to
meet Jud, the watchstander. Jud has a little job he wants help
with. I don't know what it is all about, but it has to be better

than scrubbing deck, which is what Freeman and Pop are now
doing and what we have been working at for the past three and
a half hours.

Jud isn't there, but Bullrider, the other watchstander, is. I
asked Bullrider where Jud was. He said something that
sounded like "Uh," which I took to mean he didn't know. And
if Bullrider doesn't know, no one does, and since I was sent to
meet Jud, the best way to find him is to let him find me. I
cross to the small table in the corner and pour a paper demi-
tasse cup of coffee, sit down in the barge engineer's swivel
chair, and wait.

"Six and three-quarter pounds. Can you believe it? Look.
That's his daddy when he wasn't much older." Dave has hauled
out his wallet and is spreading a small family album out on the
desk where Bullrider is working.

I get up and come over for a look. Dave doesn't appear old
enough to have grandchildren. I have to remind myself that he
has already spent twenty-three years in the navy.

". . . and there he is two years ago and that there's my other
boy. He's in high school yet. He's a big one . . . six-two, and I'll
wager he's two hundred and fifty and still growing. You think
you could get him to play football? All he cares about's that
camper of his. He's real good with wiring, you know, and he's
got—"

"Move them things, hey," Bullrider grumbles. "I gotta get
this here work done."

Bullrider is in the middle of some intricate calculations. In
front of him he has a calculator, an assortment of figures, and
a log with empty spaces waiting for his entries. He is attacking
the job with the same physical intensity he presumably attacks
bulls. (I say, "presumably" because to my knowledge the only
person who has seen Bullrider ride is Bullrider. There are those
on board who claim he is pure "bull": "I could whip his ass and
eat breakfast at the same time," Hippy says.) He is not going to
let the figures throw him, if only he can keep up his concentra-

tion. Knowing the fragility of that faculty, Dave quickly picks up his photographs and settles back into the easy chair to stare at them.

I turn away and study the multi-colored panel which sits upright in the middle of the room. Like Bullrider's figures, the panel does not make immediate sense. It is a computer of sorts and the panel shows an outline of the pontoons and the cross tanks which connect them. Red and green buttons are set in an orderly fashion about the outline and beside each combination button is a speedometer-like dial which constantly records the amount of ballast in the separate tanks. By pushing some buttons green and others red it is possible to transfer ballast from one compartment to another and thereby re-balance the rig or cause it to tip, as the case may be. For instance, there are times when we are scrubbing deck and the water refuses to run off. With a friendly word from the crane operator to the watchstander, the rig can be tipped enough to allow us to squeegee the water over the edge.

For those reasons alone ballast control plays a critical role on the rig. It is also a good place to come for a *Playboy* or a *Penthouse.* For aside from counting up figures and tipping rigs and occasionally fixing a pump, there isn't much else to do. Ballast control is a dead end on one's climb up the rig's hierarchy. It is a haven for ambitionless roustabouts.

"There you are, man. Where you been?"

"Enjoying Bullrider's company."

"Let me get a cup of coffee and we'll get out of here."

I really don't know what a bullrider is supposed to look like, but if I were a bull, I would rather tangle with Bullrider than Jud. This guy just might be able to last in the ring with Big Daddy. He's built like a tackle on on a small-college football team.

"Let's go, pardner. You carry these." Jud hands me a couple of butterfly wrenches and a small sledge hammer. "Tell you the truth, I don't think I'm going to need you, but the hell with it, right? It's better than cleaning deck." Jud should know. He's a former roustabout now ensconced in ballast control. Hunting is his life—and drinking.

There are six legs from the rig floor to the pontoons. The four in the corners are for ballast; there are elevators in the middle two. We descend in the port elevator which is round and white as is the shaft of the leg. There are iron rungs built into the wall of the shaft—our escape if the elevator should malfunction. The drop is one hundred feet. There is the story of the galley hand who used to climb all the way down to smoke pot in the pontoons, then climbed all the way back up when he was through. The things people will do to get high; the things they'll do when they get high.

Jud's right. There isn't anything to do except hold tools for him—and smoke cigarettes, and when he goes up to fetch a tool, I fight off the urge for a short nap by wandering around inside the pontoon.

There are some people who are scared to death of the pontoons. Hippy, for instance, thinks nothing of climbing to the crown of the rig, one hundred feet up, but will not go down to the pontoons. "You got thirty feet of water over your head. What happens if something lets go when you're down there? How you going to get out of there?" he argues. "What happens when you lose your balance on the crown?" I counter. "You get killed, man," he says, "but it's quick."

It is not a place for the claustrophobic. It is like a catacomb, a long, long, meandering tube about six and a half feet in diameter painted pure white but giving off a yellowish tint in the half-light. It is spotless, yet chilly and absolutely silent. You do not expect to find another living thing, yet you walk braced, prey to the imagination. I'm not unhappy to hear the hum of the elevator descending.

"I can't find that thing I needed, so the hell with it. It's almost time to eat. Let's have a smoke and get outa here. I'm not supposed to be working this shift anyway."

That's fine with me. What a tour so far! Last night, there had been an indian behind every tree; tonight, I'm dying of boredom. I should be thankful. Maybe the elevator won't work. I peer up the hundred-foot shaft. Damn, I think, that galley hand must have wanted his dope badly.

II

There are all kinds of jobs on an oil rig, but for those of us doing them, they can be broken down into good, bad, and what-the-hell. Of course, what might be classified 'good' by one may be considered 'bad' by another. For instance, I enjoy unloading supply boats for precisely the reason Freeman loathes it. It's hazardous work. Not that I am any more courageous than Freeman or have a lesser sense of peril, it's just that I hate tedium and Freeman doesn't. For his part, Freeman likes scrubbing deck. Its lack of demand suits his enthusiasm for hard work. I detest it and consequently am not as conscientious as Freeman. Pop has a close relationship with cleanliness, so that for him scrubbing deck is a good job. As an obedient employee he doesn't mind unloading supply boats although given the choice he would as soon remain on deck. As an old logger, what he really likes is putting drill pipe or casing on the catwalk. There he gets to use some of his store of hitches—albeit his choice is usually reduced to the clove hitch. Handling drill pipe falls into the 'what-the-hell' category for Freeman and me. It's dangerous but only routinely so, a break from whatever else we may be doing at the time.

If there is any job Pop dislikes on the rig, it is working in the mud room. He finds it boring. Oddly enough, for me it ranks with unloading supply boats although I agree with Pop, it can be monotonous. Freeman is rarely told to work there. The one time he was sent he was found seated on the fork lift, slumped over the steering wheel, sound asleep. The appeal for me is the

people I meet there. So, it is fine with me when Lew comes up
on the deck and says,

"Why don't you go on down and help the derrickman."

As I pass by the port crane I look up to my left at the shale
shaker room. There's Hippy in the far corner with his orange
cups and his scale, taking his weights and measurements of the
mud which has come up from the hold below. His dark blue
coveralls are relatively clean, in marked contrast with Bear's
gray ones which, like his pink, bearded face, is splattered with
mud. Given his low rung on the roughneck's ladder, Bear has
been assigned the obnoxious job of washing the "gumbo" off the
shaker screens.

The shale shaker room is the backstretch for the mud
which has circulated from the mud pits, down the drill pipe,
around the bit for lubrication, and back up the hole, carrying
with it the rock chips and sediment that have been drilled
through. The mud pours over a pair of shakers, the liquid seep-
ing through jiggling screens and flowing back to the mud pits,
the chips and sediment remaining above to be jounced along
until they fall off the edge and collect in heaps in front of the
shakers. Here they wait to be washed down the open trap or to
be sampled by the oil company's geologists. Sometimes the
sediment is primarily clay and collects in a thick paste on the
screens, plugging the mesh. This is called "gumbo" and has to
be washed off by hose. It is not a sought-after job and has fallen
to a grim-faced Bear.

I duck down the stairwell by the personnel basket, follow
the port corridor to the door marked "Sub Sea International,"
make my way past the diving apparatus, and enter the mud
storage room. Before me are two rows of huge, conical vats with
an aisle between them. Those to the left are filled with barite;
the ones on the right, with dry cement.

I turn left from the mud storage room into the mud pump
room and am assailed by a hammering roar from the two mas-

sive, deep green National pumps, their pistons driving the mud from the mud pits down into the hole nearly a mile below. I continue past the pumps and enter the first of two mud rooms. In effect there is only one mud room, a cavernous space divided in half by a partition wide enough to wheel a fork lift through. Both rooms are piled high with pallets of chemicals. The forward room also contains the Halliburton cement pump which stands out from everything else in the place because it is bright red and appears to have never been used. Which is close to the truth. It is used to cement the casing in place in the hole, to plug a drilled dry hole which is being abandoned, and on occasion to fill a "wash out" when one occurs down hole. Halliburton maintains its own man on the rig full-time and he earns his ten dollars an hour by painting and repainting and washing down the idle pump.

The mud room is a mess—as usual. Empty chemical sacks litter the floor which itself is covered with spilt chemicals turned sticky and soggy from the water that leaks out of the small pumps along the floor. Broken and whole skids stand up against the wall. Full and half pallets rest at odd angles to each other, jammed in any readily available space, their bags knocked askew, some torn from having been too wide for the narrow space they were forced into. And like an old chicken barn, there is a thin film of white dust covering everything it can settle onto.

The back half is where the mud work is done. Dominating the room is a single white vat, like, but smaller than, those in the mud storage room. It is hung from the inside wall and contains barite which has been pumped from the larger storage tanks. Directly beneath the vat is the hopper with two wide holes cut in it and funnels running out below to a large, white, bespattered pipe. On the bottom of the vat is a lever. The lever releases the barite, which rushes into the left hole and into the flowing mud in the pipe on the floor. The right hole is reserved for adding such chemicals as have been prescribed by the mud engineer and the derrickman. My job will be to work the lever.

"Mess, hunh?" Clayton, the derrickman, says in his soft, deep, Arkansan voice and smiles gently through his full, trimmed beard.

Few things bother Clayton. He doesn't appear to know words like: anxious, harassed, frustrated, frightened. A job is a job. He gives no greater value to being swung aloft ninety feet to his platform in the derrick than he does to weighing mud or measuring viscosity in the mud pit room which is up a short flight of stairs and behind the mud room vat. He likes to laugh. He even appreciates Pop's jokes. You have to know him well before he will let you know that inside that long, lean woodsman frame he is carrying a powerful sadness. His wife has walked out on him. He says she wants a little more excitement. He has plans to quit and open up a hardware store back home in Camden. She can be his partner. He doesn't know whether that will be enough. Each night before he goes to bed he reads the Bible.

"John there's got the chemicals. Why don't you take the barite. Keep it half full for a while."

I look over at the hopper. There's a new roughneck I haven't seen before.

"John?"

"I think that's his name."

One of the minor benefits of working in the mud room is you get to take off your hard hat and gloves. I do—with pleasure. I respect the need for both but I find them cumbersome, especially the hard hat. I feel the same relief a football player does when he comes from the field and can yank off his helmet.

"I've got it," I say to the new roughneck who has been trying to mix two bags of chemicals in one hole while keeping up the flow of barite in the other.

"How you doing, brother?" he asks, stepping aside to let me come along.

I haven't heard anyone called "brother" in a long time: pardner—hoss—man—babe, yes, but not "brother."

I want to say "Real good," because it seems to fit with "brother." Instead I give him a noncommittal, "Okay. How you doing?" to which he answers "Real good," and picks up one of the bags, shakes what's left into the hole, heaves it onto the floor behind him, and leaves to fetch another.

Working the lever and pouring the barite requires a delicate touch which can be learned in about five minutes. The key is maintaining a steady flow while not letting the stuff back up in the hole which, being shaped like a funnel, tends to clog when there is too much pressure from above. The same occurs in the vat. Frequently, the vat stops flowing; you open the gate wider, and the barite lets go and pours full force to the hole, spilling over the hopper and down on the wet floor. The fun in working the lever is that you get to bang the vat with a piece of pipe to loosen the barite sticking to the wall inside. The degree of resonance also tells you how full the vat is and how close you are to having to fill it again. Banging the vat is one of the few opportunities we get to personally add to the ubiquitous clamor of the rig. No matter who is handling the lever, he always seems to bang the vat more often than is really necessary.

"Name's John N—" I shake my head. His last name was drowned out by the pumps.

"John Nason," he yells. I introduce myself. We shake hands. He has a strong grip. I can see why. He has arms like Bullrider's with shoulders and neck to match. He's not as tall, nor does he have Bullrider's gut. He could be a weight-lifter or maybe a 168-pound wrestler. It turns out he was a middle linebacker.

"Where you from, brother?"

"New Orleans." I find it easier to say that, as it is partially the truth, than to come right out and admit to being a Yankee. "You?"

"New Orleans." He pronounces it "N' Orlins," the way half the natives do, the other half saying "New Orlee-ans." No one says "New Orleens" except Yankees and Country Western singers who can't fit "N'Orlins" into the beat.

"You a teacher? A student?" he asks.

"Just a roustabout," I answer, somewhat annoyed that anyone should have thought otherwise.

"That's funny," he says. "You have a professorial look about
you."

Professorial?! Now who's sounding like a roughneck?

"It's the pipe," he explains. The pipe. Bear chews tobacco,
Pop has his snuff, and I smoke a pipe. It's what we do. I'd
established that eccentricity the first day on the rig. I'd forgotten that pipe smoking is more of a rarity amongst roustabouts
than amongst teachers.

"You don't come from N'Orlins," he states.

The hell with it. "Boston," I state.

"Connecticut," he states, "Living in N'Orlins . . . You got
wheels?" I nod. "That's good, brother. Give me a ride back from
the bank?"

"Why not," I answer. Another Yankee, I think to myself,
and this poor bastard has to work directly under Clyde. He's
going to need all the help I can give him. "Be glad to."

We work on in the dinning silence. Mine is the compara-
tively easier job although the degree of difficulty in John's lies
in his having to lug the sacks of chemicals from the pallets to
the platform. For me, time is measured not in minutes but in
how long it takes to empty my vat, to fill it up again, to empty.
. . . John is on a more precise schedule. He is mixing Carbonex
and caustic soda at a prescribed rate—one bag of Carbonex
every eight minutes, one bag of caustic soda every twelve minutes. The Carbonex, a light, black powder, is designed to thin
out the mud; the caustic soda, to make the chemicals in the mud
work. The Carbonex is a noisome substance. It wants to blow all
over the hopper into our eyes. After a short time we both look
like miners. The caustic soda can be dangerous. There is a
warning clearly marked on the sack listing all the things one
must and must not do when handling it. It emphasizes we must
wear glasses to protect our eyes, gloves and heavy clothing for
the skin. To dramatize the message there is a line drawing of

a drop of caustic soda landing on an arm. There is a neat, V-shaped hole in the arm. The first time I worked with caustic soda Clayton said to me in his quiet, unexcited way: "Better watch this stuff, it'll burn the shit out of you." There are no eye protectors available. Most of us don't bother with the gloves and it is usually hot enough in the room that we have our sleeves rolled up. We pay obeisance to the danger by warning one another that "it'll burn the shit out of you."

"Who's that?" John asks. I look around. It's Freeman, leaning against the partition watching us.

"That's Freeman. You haven't met him yet, have you?" I answer. "Must be on his way to crap."

Apparently not because for the next ten minutes Freeman continues to stand there, idly gazing around the mud room. The only time he moves is when John walks by him to fetch a bag of caustic soda. The first time John walked around him, the next time he made Freeman move.

"Does he always work that hard?" John asks.

"Freeman's the only person I know who can do nothing for eleven and a half hours and have the nerve to say, 'We got half an hour to go, why don't we fuck off,' " I explain. "The funny thing is, he can be a good hand when he wants," I add. Which is true. The guy is lean but very strong and when pushed can lift twice as much as either Pop or I can. The problem lies in the pushing. "The thing about Freeman is, he thinks he's working. He gets all red-assed when you tell him otherwise."

John looks over at Freeman, who is now picking his teeth, and shrugs. "That's his business," John says. "If he can get away with it, more power to him. Just as long as he doesn't get in the way."

Still, I wonder what Freeman is doing down here, and when I go to fill the vat again I ask him where Pop is.

"Hunh?" Freeman says. "Ol' Pop washin' his clo'."

I don't have to ask whether Lew has called the deck washing off. He hasn't. Rather, Freeman is lonely. He has come searching for company. Some weeks later Freeman will be forced to spend a full twelve-hour shift down in the starboard pontoon watching for leaks while the rig is under tow. An ideal job for him, you would think—twelve hours of just sitting. Freeman nearly went crazy, not from the boredom, but from the loneliness. "There weren't nobody there." Whenever he could no longer stand it, he would call Bullrider on the bitch box, ostensibly to report there were no leaks, in reality, to hear another human voice, even if all that voice ever said was, "Awright."

Obviously our presence is enough. Freeman does not feel any urge to grab a broom and sweep up or to stack the empty sacks. It's not our job to tell him to. Pop and I have tried in the past, but all we ever got was a snappy "Hunh?" Freeman knows who his boss is and he is fully aware the boss isn't in sight.

So when the two pallets of Aquagel suddenly let go and tip over, spilling bags helter-skelter on the floor, I keep my mouth shut and wait to see what happens. I'm working, John's working, so is Clayton. The only person not working is Freeman and it is as though nothing has happened. Except that he has left his leaning post and is now sitting on the fork lift picking his teeth.

"Hey, Freeman, can you start stacking those bags up again, brother?" John says, as he passes by him with a bag of Carbonex.

"Hunh?"

"Those bags that fell over. The unbroken ones got to be stacked up again," John says.

Freeman just looks at him with a "who's this guy?" frown on his face. John stares back at him, then glances at me, incredulous, and turns his back to cut the sack and continue pouring the chemical into the hole.

"What's with this guy?" John mumbles to me. I shrug.

"It's Freeman. There's no way in hell you're going to get him to pick those up unless the crane operator tells him to."

"The hell with him," John concludes and begins clearing away the broken sacks and restacking the Aquagel on the skid. With one eye on the flow of barite, I join him. Unlike the other chemicals which weigh around seventy-five pounds a sack, Aquagel weighs one hundred pounds. There is an old oil field trick played on new roustabouts or roughnecks who arrive with an exaggerated swagger. Someone says, "If you're so strong, I'll bet you can't hold a sack of Aquagel over your head for ten seconds." This is a macho challenge difficult to resist. The new hand grabs the sack, presses it over his head, stands there with a strained but cocky smile on his face . . . and behind him someone draws a knife and neatly slits the sack down the middle, quickly draining the contents over the hero, covering him instantly, inside and out, with gelatinous muck. In such simple ways are heroes reduced to working hands.

One hundred pounds is heavy. Two hundred pounds is heavier. Three hundred pounds begins to dampen one's sense of humor, especially toward the end of a twelve-hour shift. John is tenacious, almost viscious as he grabs sacks and carries them over to the waiting skid. His silence is razor sharp. There is no reason why we should have to do this, we have our jobs, but it's there to be done.

We do it. With a vengeance. Freeman is our goad. Neither John nor I have said as much to each other, but we are going to show Freeman how people really work. We are going to shame the bastard off his butt. Afterall, it has to be impossible to just sit and watch your mates, one of whom is drawing the same pay for the same hours, working this hard and not share at least a part of the load. Not a chance. The sonuvabitch is still on the fork lift watching us. No, not watching, staring vacantly, trying of all things to stay awake.

"Did you see that?" John grumbles as we heave another sack up on the growing stack.

"The yawn?" I ask, rubbing a Carbonex-black arm across my sweating forehead.

"A yawn. I can't believe this guy," John says and abruptly walks over to Freeman and with excrutiating politeness says, "Freeman, why don't you grab that shovel and start filling up these empty bags with the loose gel."

"Hunh?" says Freeman, startled.

"The gel. Shovel it," John snaps.

"Leave it for the day shift," Freeman suggests. He doesn't move. He glares down at John. "Ain't more'n an hour 'til quitting time. No sense us breaking our backs."

"Get off'n that thing and start shoveling, Freeman. Give these boys a hand." That's an order Freeman understands. From Lew, who, unnoticed by any of us, has been standing in the opening between the two rooms taking everything in.

"Ol' Freeman here don't like to grab much," Lew explains to John. "He and I going to have a little talk when this shift is over. Freeman's going to have to shake it a little more if'n he thinks he's going to stay out here, ain't that right, Freeman?"

"Hey, I'm tired, man. I don't feel good," Freeman says, slowly picking up the shovel and almost desultorily filling an empty sack.

"You boys do what you have to do. Let Ol' Freeman here clean up the rest of this stuff," Lew says, "and Freeman, I'll be comin' back so I want to see you done a good job," he adds and leaves.

I'm content to go back to the barite and leave Freeman alone. John isn't. He is a hard worker. Later, Grady, the toolpusher, will say, "Give me a crew of John Nasons and I'll show you one damned good crew. That boy doesn't know what hard work is, he just does it." And John can't stand watching Freeman try to pour half a shovel full of Aquagel into the empty bag which persists in flopping over and spilling.

"I'll do that," he says coldly. "Just get a broom and sweep the stuff into a pile," and he grabs the shovel out of Freeman's hands.

"What you doin', man?" Freeman asks. John doesn't answer. He's almost waiting for Freeman to say the wrong thing. Freeman knows it. He fetches a broom and starts sweeping with diligence. Strong as Freeman is, it is doubtful he could stand up long with John angry as he is.

"Okay, you men, that's it," Clayton yells from the top of the stairs at the mud pits. "We've had it for the night. Let's go wash up."

"Sounds good to me," I say, pulling down on the lever, cutting off the barite. "Let's go, John."

"Right on, brother," John answers, emptying the remainders of both the Carbonex and the caustic soda into the hole.

"What a night!" Freeman offers as we brush ourselves off and head from the mud room. "I'm so tired I could go to sleep right now."

John just looks at him.

"You'd just better not come up on the drill floor," John says. "We'll grind your ass."

7

Fourth Tour

I

Out here there is no such thing as Saturday night. In fact, there is only one known day—Wednesday. Crew change. Otherwise there are only tours: first, second, third. . . . At a certain point all concept of days is lost. They become incidental, especially at night when you work half of one and half of the next. For instance, today is Pop's fifty-fourth birthday. It's not really today, but tomorrow, that is, at midnight he will be fifty-four. It will slip up on him around lunch time and six hours later he'll go to bed to wake up twelve hours later and still have his birthday for another six hours. The way Pop figures it, he gets two days of birthday because he has awakened with a birthday feeling.

"What are you going to do to celebrate?" I ask him.

"Try not to get killed," he answers.

"Fifty-fo'. You an ol' one, ain't you, ol' Pop?" marvels Freeman from his supine position on Pop's bed.

"Old enough to teach you a thing or two," Pop growls, pressing a cigarette between his lips. "Here, sew a button on that, if you're fast enough," he says and farts.

"Ol' Pop, You a dirty ol' man," Freeman says, laughing.

"And there's another for you."

"C'mon, Pop."

"Then get off'n my bed and go to work. If you're lucky, there's a supply boat waiting just for you tonight, Freeman. You'd like that, wouldn't you?"

"Oh, c'mon, ol' man . . . fifty-fo'. That's ol! Wo-o-o!"

Pop's second birthday wish appears to be fulfilled. There is a supply boat waiting and by rights it should belong to Freeman. Guaranteed, Freeman won't like this one. It has casing on it, nearly a whole load, and the sea, which was gentle and rolling when we turned in this morning, has picked up during the day. It is sufficiently rough that the captain is not taking chances. He is laying off to port and only backing to hitch up a load, then drifting down wind away from the maelstrom set up around the rig's legs.

"It's a sonuvabitch down there," Silent Nick states tersely as he arrives back on deck via the personnel basket. It must be, for Nick to vouchsafe so much information on his own.

"Come here and look at that, Freeman," Pop calls from the railing. It is not a happy sight. The waves are rocking the *States Victory* around like a toy. Just looking at it I feel queasy. I don't want to go down. I silently pray that Lew sends Pop and Freeman. I don't care how old Pop is or how sick Freeman may become, I'll be quite content to let this one pass.

Lew joins us at the railing. "That's going to be dangerous down there, men. I want you all to be real careful. If you got to send it up one stick at a time, do it. Don't take no chances."

We all nod our heads in agreement, each of us wondering exactly which of us Lew is talking to.

". . . as soon as that other roustabout from the day crew gets through eating, I want you and him to go down [Damn! he means me.] and, Pop, you and Freeman stay here and unhitch."

Mother of God! I think to myself. *The rest of the world is out drinking, partying, having fun, and I have to spend my*

Saturday night on the back end of the States Victory *risking my life unloading casing.*

". . . and when we get that one done, we got another boat coming in with chemicals," Lew announces, apologetically, I think. As he walks past me, he takes my arm and pulls me away from the railing.

"I don't trust Freeman down there and I'm kinda scared to send Pop. I don't want to get him hurt. You understand?"

No, I think to myself. "Sure," I answer, wondering whether I should find solace in standing out from the old and the infirm.

"Okay, men, let's get to it," Lew yells from the crane.

I look at Byron, the Mississippi kid, as if for the first time. We have never worked together before. I've never had to notice how frail he is. I must out-weigh him by thirty, even forty pounds, and am a whole head taller. There is a rumor, never verified, that he lied about his age to get the job. You have to be at least twenty-one to work on deck and if he is nineteen, then I'm fifty-four. I suddenly feel fifty-four.

"Let's go," he says and steps up on the basket. I step up across from him. "Take 'er up," he yells. We both stab right arms toward the leaden sky. The basket pops off the platform and we swing out over the angry Gulf. And stop there. Eighty feet up. Swaying in the wind. Lew has snatched us off the deck before the supply boat was beneath us.

I hate it when this happens. I have a little chant, sort of a rosary, I mumble to myself: "Get me outa here . . . get me outa here." When I was first on the rig, the port crane would sometimes freeze and refuse to swing to the left. We often would hang for many long minutes being jerked up and down over the Gulf while the crane operator tried to cajole the crane into moving again. Each night I prayed to Otis, the mechanic. I'll never know whether it was the force of prayer or Otis' God-given abilities, but ultimately the crane was fixed.

"Uhhh!" Lew has just released the brake and dropped us straight down to the deck of the *States Victory.* 'Uhhh' and 'unh!' are the only overt signs I allow myself to reveal of my personal terror; the first, when the basket is suddenly dropped; the latter, when it is unexpectedly jerked upwards. Byron, I notice, is very quiet and is gripping the netting as firmly as I.

"Watch yourself," Byron advises. "This casing's been rolling around pretty bad. Like to have gotten smashed up a couple times this afternoon."

I can see why. The day crew was able to get off more than it appeared from the rig and the casing is no longer wedged together between the gunnels by its own weight. There is room for it to move around. A jury rig has been hitched up with a cable from the winch pulling the end of one pipe diagonally tight against the others. It helps. Somewhat. But forget your ankle if it becomes caught between any of these sticks of casing, each of which weighs around fifteen hundred pounds.

"I'll take the front and the winch, if you want to take the back," Byron suggests. It doesn't make any difference to me. He knows what he's doing with the winch; I can fight the washing waves. The only part that bothers me is that the casing will pass over my head, not his, and if it should break loose . . . I check my cover. I notice there is a heavy-guage water pipe running along beside the gunnel. I can duck under there.

"We'll take this one, these two, and that one. They look about the same length," Byron says, pointing to four pipes fairly close to one another. There is a difference in length sometimes of two and three feet, between the casing. We don't want to hitch up one that's too long and risk pulling the clamps loose from the others. The danger is not that one stick will break free and drop straight down. That could be all right. It might bounce around, but it could be avoided. What happens is that the clamp usually comes off one end; the other holds fast and acts like a slingshot shooting the pipe down the deck. If you are in its line, you will get killed or at the least badly hurt.

"Here it comes. Let's go," Byron yells.

Sweeping down over the stern are the clamps, four to an end of the heavy double sling, tags lines fixed to each end. I quickly but cautiously go astern and grab hold of both tag lines, keeping an eye on the waves that are breaking toward me. The captain has maneuvered the boat in by the rig and into the turbulence. The rocking is severe. I have hold of the tag lines, but the boat isn't far enough under the crane. With the kind of weight about to be hoisted, Lew has to be careful how far he booms out. Cranes have toppled off the side of the rig because the crane operator has let the boom fall too far. Booms have been known to let go from the excessive weight and fall down on the deck of the supply boat below. It has happened on *45*.

"Here you go." I pass Byron his set of clamps. He rapidly duck walks up the casing to his end. I wait a second, checking to make certain we are hitching to the same casing. If we don't we could have two arrows, one at each of us.

We are having trouble keeping the clamps on. The clamps are forged steel pieces that resemble a hand held vertically, the thumb and first two fingers together, the last two fingers together, and a groove in-between. The groove fits over the rim of the casing. When the sling is drawn tight, the opposing clamps pull toward each other, the casing keeping them apart. There are four sets of clamps. They have to be put in place separately and held there until each set is on. With the rolling of the casing and the jerking of the sling as the boat dips and heaves, the clamps want to fall out or slide to the side. We have to use our knees and hands to hold them in place.

"You ready?" I yell.

"Ready," Byron yells back. "Take it up."

I take one hand off a clamp, point upwards and quickly put it back on the clamp.

Lew doesn't fool around. The slings tighten and he snatches the load out of our hands. We both react as though we are about to be run over by a car. We wait for the last second to see which way the casing is going to swing, ready to go left or right. The second it is over my head I turn for the cover by the gunnel, keeping a ready eye on the casing as it turns round and round on its way up and toward the rig, the warning: "Always watch

the load," running through my head. I duck under the water
pipe and look out like a mouse checking the cat. A wave dashes
in in search of me. It finds me helpless and washes me over from
head to toe. I curse violently and then retreat up the deck to
the winch to wait with Byron.

"Fun, hunh," Byron says as we watch Lew start to bring the
load of casing over the railing by the day office.

"How'd you get roped into coming back down?" I ask.

"What?"

We are both intrigued by Pop's antics on the rig. There is
too much noise to hear the yelling, but his arms are waving
around like a rush hour policeman's. Freeman is out of sight.
God knows what he is doing to work Pop up. It is a delicate
moment. The casing has to be nosed between the crane and the
day office and the crane can be boomed up just so far. The tag
lines are not long enough to allow the load to be carried over
the tops of the aerials on the day office, and to let that much
weight swing free is risky. Consequently, someone, evidently
Freeman, has to pick up one tag line and start pulling his end
around so that the sticks are parallel to the day office. As the
load swings, Pop must grab the second tag line and hold the load
back, keeping it in parallel. Then it must be inched along past
the office until it can be turned at right angles and set down on
the deck. It is a matter of pride not to hit the day office. To do
so would be tantamount to announcing incompetence to the
toolpusher.

"See that bent rail up yonder?" Byron asks. "Jake hit the
day office this afternoon. Knocked two pipes loose. Did that
railing in real good." It sure did. The top railing is nearly bent
down to the second.

"I still want to know how they got you down here," I per-
sist.

"Money," he flatly admits, "and anyway, no one else
wanted to." Byron means overtime. He is now making $7.05 an
hour and there may be three hours of work here.

"You have to be hungry," I say.

"My wife and me, we fixing to build us a little house 'n we gotta git us up some money."

"How long have you been married?" I ask. Next he's going to tell me he also has children.

"Round about five years, I reckon," He notices my disbelief. "Sure. Got us a young 'un, too . . . and ol' Dwayne, there, he married, too, and his wife's got a little 'un coming real soon now."

"How old are you, Byron?" I ask bluntly.

"Twenty-three . . . Dwayne, he's twenty-four. Yah, suh, my Daddy, he give all us kids about twenty acres a his land. Purty land, pardner, and we gonna git us a house on it.

"You think you can make enough out here to do that?"

"No, suh, but I got a job my other week. Mechanic in a garage. And I help my Daddy felling and hauling trees for the pulp mills. We're gittin it together. . . . Hey, let's go git this one." Inside that toothpick stands a tough kid.

We're getting them. It's a mighty game of pick-up sticks with one rule change. It is all right if the other sticks move, only not too much. I don't know how many we've sent up. I lost track after the first four. The answer may well be in miles. The only "how many" that interests me at this point is the number left. It seems that the more we get off, the more dangerous the work becomes. There's more room for them to roll in and less weight to hold them back.

About fifteen minutes ago our wedging system let go as the boat fell broadside to the waves. In a couple of seconds over twenty thousand pounds rolled abruptly to starboard, crashing into the iron gunnels, canting the boat that much further . . . and racing pell-mell after each other to port, giving the illusion the boat was working through swells twice the ten feet already harassing her: back and forth in seeming perpetuity.

"Throw a chain in there," yelled one of the deck hands.

"Your ass," I yelled back for both Byron and myself. At times like that one is very clear on which job belongs to whom.

Getting the casing off the deck is ours, keeping it on the deck is the deckhands.

"Here, grab the end of this," the deck hand shouted at me as he ran aft with a line of heavy chain over his shoulder. Caught up in his action I picked up an end and followed him. The casing was jumbled all to starboard, perched and ready for its sweep back to port. We stood like the fools of the streets of Pomplona, waiting for the bulls. As they came toward us, we threw the line of chain in their path and ran. And the casing, all twenty thousand pounds of it, stopped dead.

"Good trick," I mumbled begrudgingly.

"Now you know how to do it yourself," he grumbled and went back into the warm comfort of the galley.

There are three more sticks to go. We are both soaking wet, cold, and tired—and very efficient. And silent. We want out of here. All of which added together makes me exceptionally nervous about the remaining casing. They are all the same length; there is no way we can be hurt unless we are sloppy. The job is almost over, and I have what I call my 'last run down the hill' premonition. That's the one you get hurt on.

"Be careful on this one," I say to Byron as I hand him his set of clamps. He smiles for the first time all night. I can't tell whether he's having similar thoughts.

We hitch up. Duck soup. He waves to me. I signal Lew. The casing rises . . . and comes right back at me, like a spear. I'm so surprised by my prescience for a moment I can't move. At the last second I dive to the side so that instead of being smashed directly in the face, I receive the blow on my hard hat which is ripped right off my head, breaking the string I have around my chin to keep the hat from being blown off. The force sends me into the gunnel. My breath is knocked out, but only for a couple of seconds. I reach up and touch my forehead, then look at my hand, concluding that I still have a forehead and that not a drop of blood has been spilled.

"Your hat," Byron yells. I look around. A wave is washing

it aft. I leap for it and dig into the wooden deck with my boots to keep from flowing overboard. There is now not a spot on me that isn't soaked.

"You lucky, hoss," Byron says matter-of-factly as I regain my feet and we move to the center of the deck to await the personnel basket.

"You're the lucky one," I answer. "I've got another boat to unload. You're going to bed."

"Maybe," he says. "Maybe not."

II

There's the story about the man found slapping himself with a stick. Someone asks him why he's doing it. "Because it feels so good when I stop," he answers.

That's how I feel as I make my way back on deck. I've had a big hot lunch of veal cutlets, rice with Tabasco-spiced gravy, corn, salad, strawberry shortcake, three cups of "dark" coffee, and two leisurely cigarettes. I have on a clean, dry set of clothes, and I'm ready to let somebody else do some work. The truth is, I'm feeling smug, quite pleased with myself. I handled my end of a dangerous job, suffered the thrill of nearly being badly hurt, can now talk about it with as much relish and hyperbole as I dare, and, as a frosting, I have just shut Clyde, the driller, up at lunch.

Now that I think about it, shutting Clyde up may not have been a wise thing to do for Clyde is like a weasel, soft and silky, nice to look at, but unpredictable and capable of malevolence when cornered. At thirty-four years old he has been a driller for only a year and already his crew of roughnecks has turned over at least three times. He is dogmatic about his likes and dislikes. He detested the Old Man for his independence and Jack for his college education. He likes Bear because he is a full-blown good ol' boy and Hippy because he is a sycophant. And he despises Yankees because . . . well, tonight's exchange explains it best.

By the time I had changed, made it to the galley, and

gotten my food, there was only one seat left at the table, right next to Clyde. As a rule I avoid sitting near him, knowing how uncomfortable my presence makes him and how rancorous he is when made uncomfortable. This time I had no choice and I set myself to be as unobtrusive as possible.

We ate in silence. There was some conversation down the table. I kept to my meal and listened, trying not to pay attention to Clyde's occasional sidelong glances at my plate. As we were eating much the same thing in equivalent amounts, he evidently couldn't find anything wrong there. As I had changed out of my "blue uniform"—blue shirt with blue pants—I had robbed him of his favorite jest: "Next week I'll be dressed in gray—'Confederate gray.'" Since I wasn't saying anything, I wasn't talking "funny," so he couldn't say "I hope you got your passport in order 'case you ever pass into Mississippi." Clyde was suffering; he couldn't find an opening. Still, I could sense from his silence he was looking. When he put his fork and knife down by his plate and looked across the table at Hippy, I guessed he'd found one.

"You know, Hippy," he began, "It's a Godamned shame the North had to win the War."

"Well, Clyde," I said, also looking at Hippy, "You can't say we didn't give you boys a chance."

Everybody laughed. So did I. Even Clyde had to. It wasn't so funny but it was the right thing to say. It shut him up, he didn't have a response. All he could say was, "Fuck you," but I have the feeling I may be made to pay later. Or John Nason, the new Yankee roughneck, will.

My euphoria—if that is what it is—is abruptly disolved by the sight of the next supply boat and Lew's smile. I look around for Byron and there he is. Here we go again. I notice Freeman standing next to Pop. They are a team tonight. I want to get angry at both of them, but I know Pop isn't shirking. He'd go down without a complaint, if he were asked. I contrast big, muscular Freeman with slight, almost concave Byron, who is now nearing his twentieth hour on deck, and I focus my anger at Freeman.

"You ready?" I ask Byron. I feel like apologizing to him. He looks very tired. He shouldn't be going down to work this boat. It has pallets of chemicals on board. That means using the bars, which are lethal on the rig floor and more so on a tossing supply boat. It's not fair to send a man who has been working for twenty hours down under these conditions. But the Mississippi kid is tough. He won't complain. He nods that he's ready. Down we go.

This is going to be impossible. There must be fifteen pallets on deck and they have all been wedged together by the tossing of the boat. There are sheets of metal and various heavy tools, steel crates, all tied down with chain. None of them belong to 45, so we can't move them to get at the pallets. It's a dog's breakfast, and all Byron and I can do is stare, each of us yearning for casing.

And here come the bars, dropping down on us like daggers. I have no idea where to start.

"Think we can get them in there?" I ask Byron.

"Can try," he answers dubiously.

There seems to be enough room to angle the bars under one of the pallets. Whether we can attach the chains at the other side is another question. If we can, will we be able to hold the bars apart so that when the lift comes, the bars won't slide together and tip the pallet over.

"It's our only chance."

We each snag a flopping bar. We don't even bother to defend ourselves against the chains. They flap at our heads, banging naggingly at our hard hats. One whacks my shoulder. I signal Lew to drop the whole thing down which puts the headache ball just above our heads. It lands on top of the pallet and lies there, threatening to drop off on us. We are already soaked through again. Our hands are numb. I sense three sets of eyes peering down on us, prodding us to move faster. When I look up at the rig, I am blinded by the spotlight Pop has turned on us from on the rig. He's trying to help, but I feel helpless and frustratingly incompetent, and the spotlight has made an arena of our hopeless situation.

"You got yours through?" Byron yells.

"Just about. You?"

"Almost. Can you move yours over?"

"It's caught on something."

"The other pallet. Pull it back . . ."

I don't know how, but we finally drive the bars through. There's just enough showing out the other end to give the chain a bite.

"Take it up slow," Byron yells.

"Watch yourself in there," I warn, needlessly. He's reaching in between the two pallets and trying to keep the two bars apart with his hands.

I signal to Lew. I get one chance and the signal is complicated. If he doesn't understand, Byron is going to get hurt. I point behind me and then, lifting my hand, rub my thumb, index and middle fingers together. I want Lew to take it up slowly while inching it back at the same time. I grab the chains and yank hard, bracing myself. Lew understood. The minute the pallet is free, I signal him to drop it. We adjust the bars, fix the chains on tight, and send it up. One down. We are exhausted.

"Which one now?" Byron asks. He sounds sick. I clamber over the pallets, searching for one with some working space.

"If we can move this one back without spilling it, we can get it."

"Let's try it," Byron says. He is standing in the center of the deck, knees locked like a sleeping horse. Arms akimbo. His eyes are fluttering. He looks ghostly.

"You all right?" I ask. "Want to go up?"

"Nah, I'll be okay," he says.

It's a pretty piece of crane work on Lew's part. Despite the reeling boat, he is able to ease the pallet back and give us room for the bars. Wade couldn't have done better. Two down.

We're knocking off the pallets around the edges. We are being battered and bruised by the chains, and when not by the chains, then by the edges of the sheet metal and the corners of the boxes, as we fall and trip and scramble around the pallets, worming our way into crannies and holes and spaces that foxes on the run wouldn't try. Each freed pallet is a feat. We stand there on the deck watching them rise through the spotlight and disappear into the dark beyond.

But our string of successes appears to be broken at eight. There is decidedly no way to get the rest. The pallets are so twisted and jammed into one another that the bars are blocked. We go through motions accomplishing nothing, then give up. I stand up on one of the pallets, point down at it, and raise my hands in defeat. In the penumbra of the spotlight I can see figures gesturing back at us. Like a clown I try to sweep the light out of my eyes as I rock back and forth on the pallet, squinting to see what they are signalling us to do.

"Can you see what they want?" I yell at Byron. He shakes his head.

"Go sit down. Rest," I say. "I'm going to call up and see what they want." The captain has a grin on his face when I arrive at the wheelhouse.

"You ain't never goin' get that, pardner," he announces pleasantly.

"You're right about that," I answer. "I don't think they know it, though . . . can I use your radio?"

"Over there, man."

Gulf Star 45, *this is* Dearborn 206. *Hello, Lew.*

I've never been on this end of a bitch box. I've never been on the other end either. I know enough to squeeze the button, not enough to let it go. There's no answer. I call again.

"Let go of the damned thing," the captain suggests. I do, just in time to catch the tail end of an answer.

Try that again, will you, Lew? I call.

Look, this is Dan, the toolpusher. Now what I want you to do is. . . . What follows is a set of instructions that could only be

offered by someone who isn't there. I try a meek argument. Dan
is the boss and he sounds very confident.

Do your best, he commands. I put the receiver back on its
holder and shrug as I pass the captain.

"Do your best," the captain repeats after me, adding, "but
it ain't going to work."

Now where's Byron? I can't find him anywhere. He was not
in the galley where I'd left him. He is not on deck, at least I can't
see him anywhere. I look around behind the pallets. I call for
him. No answer. My stomach flip-flops. Someone would have
seen him if he'd washed overboard. Or would they?

I am about to call out again when I see him, or rather I see
a bundle of clothes huddled over in the far corner out of the
lights where the deck meets the superstructure. It is a pathetic
sight. He is doubled over and appears to be in pain. What in
devil could he have done to himself, I think as I run over to him?

"What's the matter, man? You hurt?" I ask, resting my
hand on the nape of his neck.

He turns a pale, washed-out face to me. "Sick" is all he can
muster.

"Let's get you out of here. Think you can ride the basket?"

"What did they say?" he asks.

"To do our best," I answer.

"That's what we got to do, then," he says and wobbles to
his feet.

There is one pallet we might be able to get. There is space
for one bar. It will mean some delicate wiggling on Lew's part.
We estimate the chances of success at twenty percent—maxi-
mum—with eighty percent that the pallet will turn over. But
Dan said to do our best and this is it.

We hitch up. I move out on deck and mime for Lew all the
maneuverings he must go through. I signal "up—very slowly,"
rubbing my palms together supplicatingly. Byron is wedged
behind the pallet. There is no room for error. I join him. We put
our backs to the load, brace our feet half way up the adjoining

pallet and push, not so much to move the pallet as to keep the
sacks of chemicals from cascading down.

It's hard to believe but it is working. The pallet is moving
out. There is almost room to stick the other bar in . . . and the
boat drops down while Lew is lifting up.

"Look out," Byron yells. We dive in opposite directions, I
toward an open deck, he between two other pallets, and the
entire load tips over sending three thousand pounds of chemicals into the space we just vacated.

"That's it. That's our best," I declare. Byron can't say anything. He is standing by the overturned load, vomiting. I look
up into the spotlight, point at Byron, and wave my arms back
and forth. We are done. No more. Get us out of here. Do it your
Goddamned selves. I unhitch the bar so that Lew can lift it out
and send down the personnel basket.

"Good try, men," Dan, the toolpusher, says. "Go down and
warm up. Change your clothes and have a cup of coffee . . . you,
go to bed. We'll get your buddy up here."

Byron doesn't wait for another invitation. It is one weary
roustabout who drags his way across the deck to the far stairwell. I hope we don't see him at shift change which is less than
three hours away. I hope his wife enjoys that house.

"Tough little kid, ain't he," Pop says.

"He's a good hand," Dan, the toolpusher, adds, the highest
compliment that can be paid in the oil field.

Shivering, I disappear below for that cup of coffee—and a
long, leisurely cigarette break.

Rope is the solution. It was Pop's idea, but Lew has taken
credit for it, acknowledging Pop's contribution by sending him
down on the supply boat with Dwayne, the other Mississippi
kid, to direct the hitching up.

For weeks Pop has been agitating for ropes to replace the
bars. Survival has been his premise. "Somebody's going to get
killed by them bars," he has pointed out correctly. "Why don't

they just make a couple of big loops out of that two-inch line
they got below and just loop it around the skids. That's the way
we used to do it back in the state of Washington." Pop has had
trouble adjusting from the ways "it is always done" in the state
of Washington to the ways it has always been done in the oil
fields.

By the time I make it back on the rig floor there are only
four or five pallets left on the supply boat. I'm glad Byron isn't
here to see how simple the job has been made. In half an hour
they have accomplished what it took us nearly three hours to
fail at. Typically, the next time a supply boat with pallets of
chemicals arrives, we use the bars. No one knows where the
ropes are. They have vanished. Their use is never suggested
again, almost as though their effectiveness were an embarrass-
ment.

"Hey, Clyde wants to see you up on the drill floor, brother."

Pop, Freeman, and I are hiding down in the starboard
storage room, ostensibly reorganizing the stock which has been
dumped there: bolts, clamps, odd fittings, rings, gaskets, pump
parts. They appear on board. Whoever has ordered them never
seems to be around. No one knows why they are there or what
they are for so the roustabouts are ordered to "take 'em down
to the starboard storage room." There are thousands of dollars
in parts alone, just lying around, their sole purpose, it seems, to
make work for roustabouts in hiding.

"What does he want, John?" I ask, frankly surprised that
Clyde's revenge has come so quickly. John shrugs and leaves.

Ballast control, the thyristor room, the mud room may all
be vital nerve centers for the rig, but the drill floor is where the
real action takes place. As a rule, roustabouts do not get on the
drill floor often. To be sent there is like being brought off the
bench into the game. As I mount the outside stairs I am torn
between the natural excitement I always feel when I have to go

up there and a cautious concern with Clyde's command appearance.

There is nothing subtle about the drill floor. Everything about it speaks of power—sheer, brute, indisputable strength. Rising straight up a hundred feet to the crown of the derrick looms a hatch-work of bridge-sized I-beams, outlined against the black sky by tiers of incandescent lights. The derrick is built to withstand hundred-and-thirty-mile per hour winds. It has a capacity of nearly a million and a half pounds. When you look up, your jaw drops, not solely from awe but also from the steep angle of tilt needed to take in its reach. And there is the draw-works, deep green, housing miles of two-inch drill line able to haul back two hundred thousand pounds dead weight of drill pipe from the hole. In the middle of the floor is the rotary table with the Kelly, a thirty-foot hexagonal pipe, in its grasp, the Kelly stabbed into the drill pipe below, the rotary turning over a mile of pipe and grinding the bit through the shale, sand and clay—the Kelly spinning so fast the hexagonal lines are a blur. And littering the floor, pipe wrenches and sledge hammers so heavy one man can barely lift them. And crowding in from everywhere the roar of the massive machinery at work, a roar that is unexpectedly punctured by a horrifying shriek, like a factory whistle, from high up in the derrick as the stabilizers release air.

And there is Clyde, almost languidly standing before the driller's console, his hand on the shank of the clutch, his eyes fixed on the rotating Kelly. He is surrounded by dials and guages. And by the rotary table are Bear and Catfish. The rotary is covered with mud; the drill floor is awash with it. Bear is filthy, covered from head to toe. He is frowning, an orange-sized chaw pushing out his left cheek. Catfish is Augean. He has a cigarette dangling from a snarling mouth. They look exhausted.

Clyde turns around, sees me, and beckons with a crook of his finger. With an obedience which strikes me as nearly fawning as Pop's I walk over and stand near him. He lets me stand there, turning his back on me, and continues with his drilling, his left hand on a lever which rises from the floor, his eyes bouncing between the dials on the console and the visible length of the spinning Kelly. To his left is a box with a small window in which numbers race, clicking off the depth being drilled. At this point they read: 6097, or over a mile of hole so far. The contract depth with Meta calls for eleven thousand feet. We are over half way there.

I wait. He keeps me waiting, not saying a word. I light a cigarette. Clyde sees me do it and motions for one, then nods his head he wants a light, too. We are not supposed to smoke on the drill floor when drilling in case we hit a pocket of gas. There are cigarette butts littering the floor. There is a pack protruding from Clyde's dark green coveralls.

I kill time watching the television in the dog house to my left. The ocean floor and the BOP stack are in focus. Fish swim relaxedly across the screen. There's a camera down there. By adjusting knobs on a control box nearby it can be run up and down the stack. It's the only way to spot problems at the well-head.

Finally, after about five minutes, he turns and says:

"Get me the key to the V-door."

The V-door? The V-door is that wooden chute which runs from the drill floor to the catwalk up which the drill pipe and the casing is pulled. It had never occurred to me before but perhaps that opening in the wall through which the pipe is lifted is also called the V-door. I was not aware there was a door there, however. When the wind is blowing, canvas is usually pulled across to shelter the men at work. Still, there are things I don't know about the rig and a V-door needing a key must be one of them.

"Where is it?" I ask ingenuously.

"If I knew I wouldn't be askin' you to fetch it now, would I?" he answers irrefutably. "Try askin' Clayton."

So I go down to the mud pits.

"Clyde wants the key to the V-door. He thinks you may have it."

"The key to the V-door? . . . Oh yeh, that key. No, I ain't seen it. Last person that had it was Catfish."

So I climb back up to the drill floor.

"Clayton says Catfish has it," I report to Clyde.

"Get it from him then," he orders.

"What you want, Perfessor?" Catfish barks.

"Key to the V-door. Clayton says you got it."

"Hell, Perfesser, what would I be doing with the key to the V-door. You best ask ol' Hippy there. He's the one knows where them things is."

So I go back down, this time to the shale shaker room. This is becoming asinine as well as tiring.

"Hippy, you got the V-door key or haven't you?"

He stares at me. Is that a glimmer of a smile?

"I ain't seen that for a couple of days now. I think it was Dan, the toolpusher, took it, but I can't be sure."

So it's over to the day office where I find Dan, the toolpusher, curled up on the couch. I start to tiptoe out, but my crashing in has awakened him.

"What you need, babe?" he mumbles not moving his head off the crumpled up newspaper he is using as a pillow.

"The key to the damned V-door," I answer unapologetically. "Clyde wants it. He says Clayton's got it. Clayton says Catfish's got it. Catfish says it's Hippy. Hippy says you were the last one with it. Tell me you don't have it so I can go back to Clyde and tell him there isn't one."

"It ain't me, babe. I've never seen it. Could be Grady has it. He is coming on duty in half an hour so he's probably up. Course he might not be and he wakes up mean. Do what you want."

I'm not going anywhere near Grady this early in the morning. I've seen him at the breakfast table and I'd rather go back down on that supply boat, bars and all, than beard Grady over the V-door key.

So foot-draggingly slowly I mount the stairs to the drill floor to report the failure of my mission to Clyde.

"What?" he yells. "You couldn't find it? God Dah-yum, babe. I don't ask you to do but one simple thing and you fuck up. Can you whistle good and loud?"

Do I have to answer that? I guess I do. Clyde is the boss.

"No," I answer. Which is true.

"Never known a Yankee that could," Clyde states with utter disdain. And he turns his back. I am dismissed.

The tour is over. We are through. It has been a long night. As I am heading down the stairs John stops me.

"You never found the V-door key, did you?"

I shake my head.

"There isn't one. It's an old oil field gag. Every new man gets it put to him sooner or later. Makes you feel like an idiot, doesn't it?"

I nod.

As I pass down the corridor toward my room, I meet Clyde coming out of the washroom, still drying off from his shower. All he does is smile. He's had his revenge. I smile back, secretly wishing Sherman had marched through New Hebron, Mississippi, instead of Atlanta.

8

Fifth Tour

"No!"

"That's right. You lookin' at Pop, the crane operator."

"Freeman, Pop doesn't know the first thing about running that crane. What's Lew want to do, kill us both?"

"I don't know, man. He tell Lew, Grady said he should practice. Pop says if he can learn it good, maybe he'll go crane operator when the rig go overseas."

"I'm going back to the mud room. I don't want to be anywhere near that crane with Pop at the gears."

"Nah, give him a chance. Lew says it's Pop's birthday and he goin' let him try."

"And look where Lew is, standing right behind him. Pop isn't going to kill Lew, it's us he's going to get."

"Maybe we lucky. Maybe it's not our day to die."

"I hope to hell you're right, Freeman. I was kind of hoping to see my family again someday."

To tell the truth, dying would be something to do tonight. In contrast with last night it has been slow to the point of somnambulism. Tonight is that turning point in the week, the

177

hours which, if you can force yourself through them, you know you will make it the rest of the way, like long distance running. We have been out here for four days, this is the fifth. At 5:30 A.M., when we go off duty, we'll have just two more days. Four days doesn't sound like much until you multiply it by twelve working hours.

What have we done so far tonight? Scrubbed deck, worked in the mud room, scrubbed deck, cleaned up an oil spill in the engine room, scrubbed deck, . . . and stared out to sea—which is beautiful tonight.

For the first time since we came out, the Gulf is almost placid. The air is still; the sky is cloudless. The stars are abundant and a nearly full moon is loafing up from the horizon, laying silver sheets on the Gulf, igniting the phosphorous along the water's surface.

In the distance to the northeast—toward land—the black horizon is speckled with the bright dots of other rigs. In the daytime the slight haze clouds them from view. Sometimes on a clear day, if you squint and are patient, you can just pick out their gray outline. Last hitch, Freeman and I had a contest to see who could spot the most rigs. Freeman claimed eight, two of which I gave him on good faith. Neither of us could have guessed the twenty or more which, in the starkness of the night's relief, are glittering.

They are cozy. They give the illusion of a coast line and make the rig appear for a moment like a passing ship. They make us seem less apart from the rest of the world, more a part of a community. Yet, their pricks of light are no larger than those of the stars overhead, millions of light years away. We are not a community, we are not neighbors at all. Not one of them would be of any use if anything happened to us out here, nor would we be to them. We are all just little gray iron islands unto ourselves plunked into a great blue wasteland. But they sure are pretty at night, from a distance.

"Now, Pop, you be sure you follow our signals," Freeman yells up at Pop.

Pop waves him off, almost disdainfully. He wants to give the impression he has everything under control. Freeman is not sure. "That old man awful uptight," he mumbles.

The first thing to do is hitch the bars on. We pull them into an open space to give Pop some room when he drops the headache ball. Freeman points to them, then signals down. The diesel in the crane kicks alive, belching a blast of black smoke. All eyes focus on the tip of the boom. Painfully slowly Pop inches it overhead. Freeman clenches his fist, telling Pop to stop it right where it is. It continues to move around.

"What's that old man doing?" Freeman says and starts swinging his arms around to attract Pop's attention. But Pop is in a world of his own. His eyes are riveted on the boom. Even from where we are standing, we can see that he is so taut you could pluck a tune on him.

"Over here, Pop. Over here," Freeman yells. Slowly the boom comes back, this time stopping overhead.

"That's good, old man," Freeman says more to himself than Pop. "Now drop the ball," and again signals down.

Operating the crane is much like patting your head and rubbing your belly at the same time. One hand raises, lowers, and swings the boom, the other hand is in charge of doing the same with the cable. Sometimes the two coincide, other times they must work in opposition. Regardless, the two hands must be in communication—which Pop's are not because he is lowering the boom and not the ball so that the two are going to meet if he doesn't pull back on the boom or release the cable. There is supposed to be a safety catch on the cable which will stop everything before the ball climbs into the tip of the boom. Since the ball will not fit through the boom, the ball can be snapped off and plummet down to the deck. With its weight and momentum from that height, it will give more than a headache if it hits someone. Of course, the safety catch has been broken for a month and no one has thought it important enough to fix. With trained crane operators working, the chances of the boom and ball meeting are negligible.

"Watch out!" Freeman yells. He didn't have to. I have already ducked behind a pallet.

"Get it up again, Pop," Freeman shouts. He does, only he forgot about the ball again. This time the ball drops straight down while the boom is going up.

Now Pop is in trouble, for in moving the boom up he has swung in the cable, which has set the ball arcing backwards and forwards like a pendulum, lower and lower towards the deck.

"Get out of the way, Freeman" I shout. This is war, medieval style. There is no telling where that seventy pound cannon ball is going to go. There is no place to hide. Freeman and I are safe on one side, utterly exposed on the other.

Thunk! It lands on top of our pallet. The cable loops down onto our hard hats. We leap out and away. I grab the cable, shouting to Freeman to get the ball's hook. We hold on for dear life, yelling to Pop to stop doing anything.

But Pop is not in a listening mood, he is on the verge of running amuck. He is, as Freeman tersely expresses it, "crazy." He has started raising the cable again, jerking it out of my hands, then the ball from Freeman's. At the same time he has brought the boom around, launching the ball in a deadly arch right into the derrick with a resounding crash. It is testimony to the temper of the steel that the ball doesn't leave a mark as it rebounds and heads back toward us, dropping lower. We dive in opposite directions and the ball lands on the deck, bouncing and tumbling and coming to a stop against the side of the mud room hatch.

Freeman and I are catching on. We don't move. We don't go anywhere near the ball. We stand where we are and stare at Pop. Lew is no longer standing behind him. He has moved outside the crane and is on the walkway by the crane door; the SOB is laughing.

What else is there to do? Pop, the crane operator! If Pop wants to go to Brazil so badly, he is going to have to find another route. Hell will be long frozen over before anyone will hear over the bitch box: *Calling Pop, the crane operator.*

"Don't touch nothing," Freeman yells at him. "Leave it lay."

Gingerly, like members of a bomb squad, Freeman and I inch forward toward the ball. Freeman has his fist raised and clenched, the sign to "stop, don't do anything." Then, like cats on a bird, we pounce; I, on the cable and the ball; Freeman, on the hook.

"Drag them bars here," Freeman orders. Freeman has taken command. We hitch the bars to the hook, then with Freeman controlling the ball, I drag everything over to the nearest pallet. Freeman continues to shake his clenched fist. I steal a look at Pop. He is poised, ready to set his noncommunicating hands to pushing and pulling gears.

Unhappily, we have to have the ball lifted in order to connect the chains to the bars.

"Wai' a mina'," Freeman says and picks his way around the litter on the floor to the crane where he stands, hands on hips, waiting for Pop to acknowledge his presence.

"Ol' Pop," he yells up. "You wan' kill us? You follow directions now, Ol' Pop, or you goin' do this by yasel'."

Pop is furious. His lips are tight. I imagine his ball of a nose and the two tiny, apple cheekbones are bright red. He rises half out of his seat, leans through the window, and yells,

"Get back over there and earn your keep."

Now Freeman is infuriated. He sets himself more firmly on the flat of his feet, anchors his hands to his hips, and yells back:

"I ain't goin' get killed 'cause some ol' man thinks he know somethin' he don'. You do right, now, Pop."

What a Donnybrook! Lew is smiling, I can't help but laugh, Freeman's anger is half pretense. Only Pop has failed to find the humor in the situation, which naturally makes it all the more risible.

"Easy, Pop . . . slow now . . ." Freeman is rubbing his fingers together. Both of us are balanced on the balls of our feet like prize fighters, pushing against the pallet, guiding it to the hatch, but ready to dance away should anything happen.

"Up now . . . up . . . up . . . boom it down . . . that's good
. . . to the right . . ." Freeman is talking Pop through the lift. Pop
is too far away to hear a word. Freeman is only trying to keep
himself calm.

Miraculously, Pop is doing it right. Maybe there is some
hope for him after all. He has the pallet exactly over the hatch.
Freeman signals him to lower it. "Stand clear down there," I
yell.

"Pop! What you doin?" Freeman shouts.

Pop's success, it appears, was accidental. He has boomed
down instead of dropping the ball. The pallet has caught on the
lip of the hatch. It is tipping over. Only the chains are keeping
the sacks from spilling.

"Get It Up!" Freeman shouts. I have turned and run for the
safety of the anchor. By the time I turn around, the pallet is free
and following me.

"God Almighty!" I moan and duck as the pallet careens
overhead. I look up at the boom which is vibrating like a spear
just stuck in the ground.

Freeman is through laughing, too. He is standing straight
up like a wooden indian, arms defiantly crossed across his chest.
"You a crazy fuckin' ol' man, you know that, Pop. Get out here
where you belong," he yells. His howls have the net affect of a
fart in a wind storm. Pop is mesmerized by the vibrating boom.
His eyes are glued on it.

"Give him a chance, Freeman," I say. "It's his birthday."

"What's going on up there, brother?" John calls up from the
mud room.

"We got a cowboy on the crane." I answer. "You better stay
clear of the opening. Old Pop is clearing off the rig tonight."

"You watch me now, Pop," Freeman hollers. "Don't do
nothing till I signal you. . . . Okay, now. . . . Boom up. . . ."

And at last Pop settles the pallet inside the hatch and drops
it, full steam, to the floor below, shattering the skid, sending
sacks of chemicals in as many directions as they can go. There
is a plaintive: "God Damn," which floats meekly up from below.

Freeman and I peer over the edge of the hatch expecting to see twisted arms and legs and oozing blood. Instead, there are Clayton and John peering white faced back at us.

"Get him outa there," Clayton says.

"Please," John adds.

"Talk to the boss," I answer.

"It's Pop's birthday," Freeman explains.

"Tell him we got a little present for him when he's through," Clayton says.

Pop's career as crane operator is mercifully curtailed. By the time Freeman and I have stopped talking to the roughnecks, Pop has been relieved of the controls and is making his way down off the crane. By unspoken agreement Freeman and I do not mention the recent muddle which leaves us with little to say since we are both still shaking.

"Well," Pop says as though nothing exceptional has happened, "a little more practice and I think I'll have it. I could feel it coming back right at the end there."

"You gotta keep the boom over the ball, Pop, is all," Freeman offers generously.

"Let's get this one next," I say quickly, and signal Lew down.

Coffee breaks are masterful inventions and no one appreciates them more than Lew unless it is Bullrider or Otis, the mechanic. Now, one of the more curious qualities about the men on the rig is that they don't talk much during coffee breaks. For the most part they are too caught up with their "Vy-eeny Weeners" and crackers and then most of them are too tired, especially at the three o'clock break, to have much to say or the desire to say it. In contrast, Lew, Bullrider, and Otis look upon breaks as occasions to expatiate and each being good Southerners they have the knack. The only trouble is, each expects to have the floor and is not a little piqued when he discovers that one or both of the others is also present. Tonight, to the delight of those of us who miss not having a working television to entertain us, all three are in attendance and, better yet, are wound up tighter than clocks.

Both Otis and Bullrider were already at the table when we came in, Otis taking up his position at the kitchen end of the table, Bullrider by the coffee pots. Hank, the welder, was also there as was Dave, the electrician, and one of the divers, a giant of a man from Arkansas with a propensity for silence, a full black beard, long slicked-back hair running over his collar, and two massive arms traced with tattoos, the most impressive being a heart on his left forearm with "Martha" written in. He once admitted his pride in that tattoo. He challenged anyone to discover the "Betsy" which had preceded "Martha." The change over had cost him a fortune, he said, but it was one of the conditions under which Martha had agreed to marry him.

Otis is a natural-born story teller capable of kneading the most miniscule incident into a matter of stupefying moment. There are those on board who will not listen to him. There are others of us who are selective. Otis doesn't recognize the difference. He only needs an audience the way a car needs a battery. Once his mouth has turned over, he will talk on until he runs out of gas. He has been caught more than once chattering to himself as he walks down the corridor. The more literal minded on the rig say that Otis is not above altering a fact that gets in the way of his story, but as I am not as privy to the facts as some others are, I am more willing to suspend my disbelief and let him roll on, especially when he is launched on "Crazy Herman" and "Mad Fred." Which is what is happening tonight. What triggered him, I don't know. He was already under way when I sat down across from him.

". . . and I'll tell you, hoss, he was hopeless up there. Lord knows how Herman ever made roughneck, but he couldn't do nothing right. He was always getting pinched by the tongs or dropping pipe on hisself. Got so when they was trippin' pipe they'd send Herman for coffee 'cause nobody wanted to be there when he finally killed hisself. I don't think he'd have

stayed roughneck for long anyway but the thing that done him
in was when he was runnin' the shale shaker room and the
company man comes in and sees that we're losing mud. Now
that's part of Herman's job there to see that. I mean, when
you're losing mud, there's something wrong. 'Hey, Herman,'
the company man says, 'You're losin' mud. How long that been
goin' on?" And ol' Crazy Herman says, 'Boss, I ain't losin' mud,
I just can't figure out where it's goin.' That's when ol' John
Hauser, who was toolpusher 'fore Grady, busted him back to
roustabout. Only time I even seen that happen out here. Ol'
John didn't like him so much anyway and he tells Herman, 'You
goin' roustabout the rest of this hitch, then you goin' to the bank
and not comin' back.' Next hitch, there's Ol' Herman and you
know what? He's goin' mechanic. What Herman knows about
machines couldn't get a cup of coffee made. When ol' John
Hauser sees him standing there on deck, a big grin on his face,
he blows up. 'Pack your bags, Herman,' he says. 'Next boat in,
you're on it.'

"So the next day there's a crew boat here and Herman has
his bags packed and Hauser's right next to him to make damned
certain Herman gets on it. Well, I'll tell you, pardner, Herman
nearly doesn't make that boat. There he goes down on the
basket. He's got a big ol' hat on like the Bullrider's there and
these big fancy boots, and he's all alone going down and he can't
resist a little cowboyin'. He can't wait till the basket settles on
the boat, he's got to jump off. You know what happens when
you try jumpin' off the basket? It moves, you don't, and there
goes Herman right down into the water outa sight. There's his
hat floatin' away and there's his bags floatin'. Everything's
floatin' but Herman. He's lost for sure. The sharks got him, we
all think, when here comes Herman shootin' outa that water,
straight up and screamin' and leapin' for that boat. I'm tellin'
you, pardner, that Herman was six feet tall at least and all six
feet of him was out of the water. Even ol' John was standin'
there laughin'. And you know what Herman wants? He wants
the captain to go get his hat for him. He was crazy, that Her-

man. He couldn't do nothin' right. . . . Not like ol' Mad Fred. Now there was a good hand . . ."

Part of me wants to hear the "Mad Fred" story again. It is one of Otis' best. Mad Fred with the last name "thaaat long," who used to wear only a T-shirt in the middle of winter on the drill floor and instead of taking breaks would climb to the crown, crossbeam by crossbeam, and for fun would walk the length of the rig—on his hands ("You'd be sittin' in the day office and outside the window there'd go these feet."), who lived in his VW bus and had his checks posted to "just about every vacant warehouse from Galveston to New Orleans."

But I'm not going to be allowed to because Lew has had to sit in silence too long and he has finally picked up a cue—Mad Fred's strength—and digging into his repetoire of strong-man stories, has begun to regale us about a roughneck he'd worked with in Abu Dhabi.

"Now I'll tell you strong! I saw this guy one day pick up a set of them casing tongs and put 'em up on the catwalk all by himself . . ."

"You think that's good, there was this guy I know . . ." Now Bullrider has begun and his Hercules could pick up a truck axle and carry it across the garage.

Unfazed, Lew pushes his behemoth back into the ring with a tale about picking up two Arabs who were fighting and heaving them both at the same time clear down the corridor "and these weren't your little Arabs neither . . ."

". . . then there was this time this guy was switching motors on his truck," Bullrider continues undaunted, in fact impervious to Lew's reinforcement, an infraction which Lew will not abide and which induces him to add more feats of muscle to the improbable set he has already put forth. Which only goads Bullrider, who raises his voice above Lew's, who becomes more insistant on the unlimited strength of his man. To a point at which neither of them can possibly be listening to the other, but have turned to the rest of us for support, demanding with their volume that we take sides in this battle of proferred titans.

The diver speaks for all of us:

"Shit."

Both Lew and Bullrider stop in mid-breath.

Almost on cue over the bitch box comes:

Lew . . . Lew, the crane operator . . . crew boat coming in.

And with an evident sigh of relief, Lew stands up. "Let's go, men," he says. And the dubious heroes settle back into the dust from which they had been wrought.

*   *   *

"This one's yours, Freeman," Lew announces when we reach the rig floor. "I think your stomach can handle it." The Gulf is almost dead calm. The crew boat is scarcely moving as it idles below the rig. It's the *Sea Wolf*, the boat that had brought us out to the rig, last Wednesday.

"Looks like some Schlumberger equipment and I hear there's some milk on board," Lew continues. "We'll get the equipment first."

This should be simple, a nice break from scrubbing deck, a good way to close out the night. Carelessly, Freeman and I slip on the life jackets and don't bother to buckle them. We jump on the basket, casually loop our arms through the netting. Freeman yells, "Take it up," and like that we are whisked off the rig and out over the Gulf as though suddenly a giant bird had swooped down and carried us off.

"Lew crazy tonight," Freeman says. I can't say anything. I'm too busy trying to get my heart out of my stomach. If I didn't know better, I'd guess Lew had been drinking.

Now he drops us straight down to the boat as fast as the cable will pull out and stops us abruptly ten feet over the deck so we are bouncing like a ball at the end of a rubber band. Freeman and I are no longer being casual. We are holding on with both hands and are leaning into the basket.

On a gentle night like tonight even Pop could hit the boat with the personnel basket, but not Lew. He is being as effective as a spastic threading a needle. He is finding everything but the deck to land us on and now the frayed tag line hanging from the bottom of the basket has caught on the railing and won't come loose. He is trying everything, swinging the basket from side to side, yanking it straight up. It won't come loose. I holler for the deck hand but no one appears. The crew boat captain doesn't know what to do. He tries to back under the basket just as Lew tries to swing the basket to the side.

"Jump, Freeman," I yell. He is on the inside, I am on the outside. He has deck under him, I have water. Freeman hesitates. Finally he jumps to the deck just before the basket falls over the edge and I am now at a thirty degree angle to the water, sheer fear and what muscle I have keeping me on. Freeman takes out a knife and cuts the tag line and the basket breaks free and at last I am upright again. With Freeman's help, Lew finally lands the basket on the deck and I jump off. I have no idea where my heart is now, but I know I haven't lost it. It is pounding like a jack hammer. I have to sit down.

It is Schlumberger equipment, all right, which must mean that decisions are about to be made about the hole we are drilling. Schlumberger is an international well-logging outfit which is called in to determine the probable productivity of the well. At last count we had reached nearly 7,500 feet, still four thousand feet short of contract depth, but close enough to begin raising the question whether it is worth drilling much farther.

Just as we'd all guessed, Freeman has no trouble working on boats. The take-charge mood he first showed when Pop was on the crane has carried over and he is directing the hitching up of the equipment like an expert. Pop has always maintained that Freeman "can be a good hand, when he's a mind to." Where Freeman makes his mistake is in letting us know this.

Lew is still having his difficulties on the crane. Pop's spectacle must have rubbed off on him. He has yet to make a clean lift, a fact which has not gone unnoticed by the captain, who quickly guns the boat out from under each load.

"What's with that guy?" the captain yells down at us.

"Full moon," I yell back.

"He's a coon-ass," Freeman adds.

"He's a crazy motherfucker," the captain concludes correctly.

"Hey, ol' Freeman, you looked real smart down there, didn't he, Pop?" Lew is positively boisterous tonight. He must be smelling some land breezes no one else is.

"I think he's about ready for the supply boats. Way he was hitching up down there, you'd think he'd done it before," Pop answers, poker-faced.

"Ah got take me a leak," Freeman states.

"See if you can hold it for another hour, Freeman," Lew suggests. "Let's get this milk down to the cooks. Take your time. Don't bust nothing. We got the rest of the night."

There are twenty, five-gallon cartons stacked haphazzardly by the personnel basket. Not one of them is symmetrical any longer. Some are dripping milk; others are slightly soggy. The way they have been treated it is a wonder any of them have anything at all left in them.

We grab a carton, hoist it up on a shoulder, and parade down the deck to the day office and the stairs leading down to the port corridor and the galley. Lew said to take our time, so intuitively Pop and I fall in behind Freeman who acts as a ready-made governor. As we pass by the vent, we are met with the smell of frying bacon and sausage. In the galley the day shift has already gathered for breakfast. We drop the cartons in the storage room. As we are leaving Freeman spots a desk in the far corner and notices that under the glass top are pressed some Playmate centerfolds. He eases over and last seen was ogling them. On his way past the galley counter, Pop snitches a piece of bacon. We are in no hurry.

Back on deck we hoist another carton each and without thinking much about it retrace our way back to the galley. The milk isn't heavy, just awkward. The night is balmy. We work up a little sweat on this second run. Leave the cartons in the storage room. Return for the next load. Look around for Freeman. He's not there.

"You seen Freeman?" Pop asks as we start back with our third load. I shake my head.

He's nowhere in sight during the fourth trip . . . and the fifth. It's been a long night. We are wearying. We've begun to sweat more and we are becoming irritable.

By the seventh trip Pop has grown angry. "That guy is going to piss himself to death someday," he mumbles as we sidestep down the stairs.

In the galley we meet Lew, who is drinking a cup of coffee and chatting with Jackson, the cook.

"Freeman helping you some?" he asks, as though he knows the answer.

"Haven't seen him in some time," Pop answers. "Must be taking that leak." Pop has to be angry to have offered that much. Usually he would have covered for Freeman.

"You all take it easy. I'll find his ass," Lew declares and finishes off the coffee in a gulp, then leaves.

Pop and I are making our last trip when Lew appears on deck.

"Take those down and follow me. I want to show you something," he says.

We drop the cartons on the pile in the storage room and follow Lew down the corridor, through the diving room, the barite room, into the mud pump room to the mud room. Lew stops and points. He puts a finger to his lips and whispers:

"Look."

There, leaning against a double pallet of Wal-nut in the corner by the door leading to the starboard corridor, is Freeman. He is sound alseep. On his feet.

"Goddamn," Pop exclaims.

"Leave him be," Lew says and signalling us to wait, goes up to the mud pits. Clayton and John come down and stare disbelieving at the inert Freeman. Lew signals them to let him sleep, and the three of us return through the mud pump room and up the stairs leading out to the afterdeck and the heliport.

"We'll see how long he'll stay there," Lew says.

The answer is, as long as it takes to shower, change, and get ready to go to breakfast. Which is what Pop and I are about ready to do when the door flies open and an irate but chagrined Freeman barges in.

"Why do you do that, man? How come you don't tell me we're off?"

"Do what?" Pop asks innocently, buckling his belt, and fishing his pack of cigaretts off the chair.

"Hey, man, we a team. We got to watch out for each other."

"That's right, Freeman. We did. We knew you was a growing boy and growing boys need their sleep. Isn't that right?"

"That's right, Freeman," I answer, "At least that's what the toolpusher told us."

"Toolpusher? He see me?"

"That's right, Freeman," Pop says.

"Damn. What he say?"

"That he's going to help you pack your bags." Pop says.

"See you at breakfast, Freeman."

We push past him and start down the corridor. Suddenly Pop stops and turns around.

"You in big trouble, Freeman," he says, grinning.

"Aw, Pop," Freeman groans, catching on. "That ain't nice, ol' Pop."

" 'Ol' Pop' you in the ass," Pop says. We continue on to breakfast.

9

Sixth Tour

Mud going into the hole . . . fourteen-point-five, Clayton, the derrickman, calling from the mud pits.

Mud coming out of the hole . . . fourteen-point-five, Hippy, the motorman, calling from the shale shaker room.

It's a routine bitchbox exchange that goes on periodically throughout the day and night while the drilling is moving ahead. Every fifteen minutes, sometimes as infrequently as every half hour, the two areas compare mud weights. Those involved with the drilling—the toolpusher, company man, driller, mud engineer, geologists—can listen in and draw from the numbers what conclusions each is responsible for.

Most of the time the numbers pass in one ear and out the other, for when things are going well the numbers coincide. What went in is what comes out. Everything is in order.

But tonight is different. The numbers have been varying and the reports have been given with greater frequency. What has been going into the hole has been heavier than what is coming out. Something below is diluting the mud and that something is gas. That could be trouble. It has to be watched.

A blowout is a possibility. Everybody is a little tense, nerves are on edge. Almost everybody, that is.

Mud going into the hole . . . fourteen-point-five.
Mud coming out of the hole . . . fourteen-point-two.

James Austin Stantley, known as Grady to the oil field, Austin to his wife, James to his fellow Tennessee Walker breeders, is sitting down in the welders' room, polishing a steel rim which appears to be a cast-off part from the BOP stack. He has an idea that if he can work a good sheen onto the rim, he can somehow use it for a table in his living room back in Columbia, Mississippi. He says he has a friend who claims he'd done it and it had come out "good." What the friend didn't tell him is what to do after the rim is polished. For instance, how do you put legs on a steel rim which is only a foot and a half in diameter? "I'll get to that when I get there," Grady says.

The welders' room is buried off the port corridor almost opposite the washroom. It is a spacious room, once white, now sooty gray. It is littered with iron scraps and welding rods. There are barrels with scraps in one corner, large wall lockers in the other, and a long workbench on the far wall. The floor is striped from past paint jobs. There are random burn spots from past cutting. The acrid odor of cutting and welding lingers. It is also quiet there. From where he is sitting Grady cannot hear Clayton's report from the mud pits nor Hippy's from the shale shaker room, which is just about overhead. All he can hear is the "shh-shh" of the sandpaper which he methodically rubs back and forth along the rim.

For the first time since he arrived on the rig nearly a week ago, Grady seems happy. The simple act of sanding has transported him back home. He has only one apparent worry: Will Night's Rambler, his beloved Walker, be ready for the show in Tylertown, next Saturday. That'll be the first show he will be able to make this year. The horse hasn't been worked much. Grady also had him "cut," last winter, because he'd become ornery. He seems calm enough now, but what will he do in competition? "He's a good horse. If he behaves, I can win with him. If he don't, I'll have to get rid of him." Grady likes to win. He isn't wealthy enough to lose.

Mud going into the hole . . . fourteen-point-four.
Mud coming out of the hole . . . fourteen-point-one.

At the opposite end of the welders' room from where Grady sits stands a large metal cage, a miniature burn basket, open on the bottom, a door in one side. It is slightly rusted, but the weld spots at the joints of the framing pieces are fresh. The cage is still wet from its first coat of the anti-corrosive solution which I have been told to apply. When it dries, two layers of pure white enamel paint will be sprayed on it. When the next supply boat arrives, the cage will be carefully loaded on, taken to the 'bank' at Sabine Pass, unloaded, placed in the back of Dan, the night toolpusher's, truck, and driven home to Hazelhurst, Mississippi, so that Dan can carry his new coon dogs hunting.

It is a nice piece of craftsmanship. The welds have been ground smooth, the sharp edges slightly rounded. Dan is fortunate the rig has two such expert welders who were able to find the time to build the cage for him—and some roustabouts available to paint it. "Be careful. Don't miss anything. Do it right," Lew insisted. In the brief time I knew him, I never again heard Lew even hint at excellence.

Dan is a task master. He is also on duty, which is why Grady is in the welder's room, mud weights the furthest thing from his mind.

Mud going into the hole . . . fourteen-point-five.
Mud coming out of the hole . . . fourteen-point-zero.

Hippy is only twenty-three years old but he is going to be a toolpusher someday and that is because he loves his work and is good at it. This might come as a surprise to his high school graduating class back in Magnolia, Mississippi, and the girls at the disco clubs in Baton Rouge, where he now lives. It would certainly amaze his superior officers in the navy from which he was dishonorably discharged for, he says, "doing a little drugs." In his lighter moments, Hippy explains his love for the oil fields this way:

"Where else can you find joints thirty feet long, dope in five gallon buckets, and a pusher on every rig?" (Joints of drill pipe, graphite grease for the ends of the pipe, and toolpushers.)

In his more serious moments, he says it is the work. He loves being able to risk his life high over the moon pool; he looks forward to the times he has to climb one hundred feet into the crown of the tower; he is fascinated with the machinery and has made it his business to understand how each part works; and he looks with disdain on anybody like Jake, the crane operator, who he decides is lazy. He is smart, a good hand, ambitious, all of which should assure him a straight climb up the ladder to toolpusher and perhaps even to superintendent someday.

But Hippy will climb faster than others with similar aptitudes because, in addition, he is a "suck ass" which is the oil field's way of saying, "political." He rarely fails to sit with or near the toolpushers at the galley table. He is at Clyde, the driller's, side like an appendage. He is the first to laugh at the bosses' jokes or to offer to fetch them a cup of coffee even when he is not getting one for himself.

All this is acceptable. If Hippy wants to be a toolpusher, that's fine. There is no one else on the rig much interested in competing with him. But the line is drawn at Hippy's penchant for ratting. Hippy is an informer. That is one of the first things you learn when you come on the rig: "Say anything you want but, if you don't want it to get to the top, don't say it in front of Hippie." As a result, no one bitches much when Hippy is around, something which he seldom realizes since, when Hippy is around, he does most of the talking.

Tonight, Hippy is very intense and silent. He has one thing on his mind—mud weights and viscosity. He is standing at the sink and counter in the shale shaker room, meticulously weighing the mud with the care and concentration of a laboratory scientist. His paraphenalia is set out in order before him—one large, orange measuring cup filled with freshly scooped mud; one scale consisting of a calibrated bar, a cup on one end, a sliding weight on the other, and a tripod for balancing; and one wet, mud-foul rag.

Delicately he pours the mud into the scale cup, letting it run over the side so that all bubbles are out. Carefully he sets a cap on the cup, gently screwing it on, squeezing the mud tight. Slowly he takes control of the cup with his right hand, placing his thumb securely over the breathing hole in the top of the cap, and places it under the running faucet in the sink to his left, thoroughly washing all the outside mud off. He shakes it, getting rid of what water remains. He takes up the towel and drys it. Gently he sets it on the tripod, then taps the sliding weight along the bar until the cup and its shank are precisely in balance. He bends low, hands on knees, peers at the numbers, straightens up, walks around the end of the counter, picks up the speaker on the squawk box, and stands there, staring at his gold wrist watch, waiting . . .

Mud going into the hole . . . fourteen-point-five.
Mud coming out of the hole . . . fourteen-point-zero.

. . . and quickly unscrews the cap, places it on the counter, pours the cup of mud back in the tank, then proceeds to wash off all the paraphenalia so that it is immaculate. Even the counter is swabbed clean. And the towel is rung out and spread out over a nearby pipe to dry, another towel picked up and laid on the counter.

He checks his watch again, writes down the weight in a small ledger book he has in the back pocket of his blue coveralls, then draws out a crunched pack of Marlboros from his breast pocket, and lights up a cigarette.

"Hey, man," he says. "Why don't you go on down and fetch me a cold glass of orange drink. And a package of them crackers."

When Hippy asks you to do something, you jump. He is

going to be a toolpusher some day. Already he knows how to give orders like one.

Mud going into the hole . . . fourteen-point-five.
Mud coming out of the hole . . . thirteen-point-seven.

"The analysts" are the cutest little pair of men you've ever seen. Short, incongruously academic-looking, impeccably dressed in dark blue coveralls, THE ANALYSTS printed in bright red across the backs, they bob up and down between the shale shaker room and their own shack above the shale shaker room, rarely speaking to anyone, little white bags and a trowel in hand. They lean over the rail to the base of the shale shakers and scoop up some of the sediment the vibrating machines have shifted out of the liquid mud. Carefully, they fill their bags, draw the purse strings tight, and bob back to their shack, duck inside, and firmly shut the door behind.

These are the petroleum geologists. They are on contract to Meta, not us. We know very little about them and even less about what they do in that strange, dark blue, windowless shack. It is clearly alien to our rig. We walk by it fifty times a day and instinctively give it wide berth.

The truth is, the shack contains a large, multi-buttoned, many windowed computor designed to monitor the various data that emanate from the hole. Every foot drilled reveals new information about the sediment being worked through. The computor collects and stores this data, periodically spitting out its findings on rolls of white paper that spill off the narrow counter and run toward the floor.

The analysts collect the paper, adjust the computor's program, periodically collect soil samples which they set in a line outside the shack to be sent back to New Orleans on the next helicopter, and otherwise read to keep from falling asleep. Unlike the rest of the crew they eschew Louis L'Amour in favor of anything that even vaguely smacks of science fiction.

But tonight they are caught up in the tension and are having no trouble staying awake. They are becoming exhausted from their trek up and down the stairs. They are more silent and purposeful than ever. They give the impression that if you could get close enough to them, you would hear them mumbling: "We're late, we're late, for a very important date."

Mud going into the hole . . . fourteen-point-five.
Mud coming out of the hole . . . thirteen-point-five.

"Stand back! Stand back!"

I've never seen Dan Walton, the night toolpusher, so tense. He is showing his youth. Most of the time, he is calm, easy-going, the kind of boss who will stop what he is doing and answer a question, lend a hand if he sees one could help, explain some point about drilling. He has only two interests in life—oil drilling and coon dogs—and they consume him to the exclusion of all else in the world. But what distinguishes Dan from other bosses is that he wants you to know as much as he does, and anyone who shows the slightest curiosity has immediate access to his ear.

But not tonight. He is giving orders and he wants action yesterday, if possible. And what he is worried about is the BOP stack. According to the lights on the panel on the drill floor, one of the choke valves isn't closing properly. He has stopped all drilling. He isn't going to take any chances. We are working through gas. No one can tell how much, but then again it doesn't take much to blow back out the hole. Until that safety valve can be closed, the rig will be shut down.

The drill floor is abnormally silent. Everyone is standing at his post at attention. We are all watching Dan and Clyde as together they brainstorm through the system that should close off that valve.

"Okay," Dan says. "Try it. . . . EVERYBODY STAND BACK. If this thing don't hold, there's going to be mud all over this place."

They are going to shoot seven thousand pounds of pressure

down the hole. If the valve is shut, nothing will happen; if it is still open, all the mud below is going to stream up the line.

Freeman, Lew, and I hide behind the draw works with Bear and Catfish. Clyde ducks into the dog house. Only Dan remains out on the floor peering warily down the rotary hole. Catfish lights a cigarette.

"Hey, Catfish, there's gas on the floor." Freeman whispers.

"Fuck 'em," Catfish spits back and nods toward Clyde, who is quickly puffing down a cigarette out of Dan's sight in the dog house.

"What happens it don't work?" Freeman asks.

"Wha' do we have them divers sittin' down there eatin' for?" Catfish answers.

"Ain't nothin' comin'," Dan yells.

"Guess we got it, hunh, babe?" Clyde says.

"Run the pipe back in," Dan orders. "The rest of you all, get on back to what you were doin'."

Mud going into the hole . . . fourteen-point-five.
Mud coming out of the hole . . . thirteen-point-five.

Galen, the mud engineer, is one of those people on the rig who is there but you never see him. It seems impossible. For eighty-four of the one hundred and sixty-eight hours spent on the rig you are wandering all over it. Over seven days you have twenty-one meals and close to fourteen coffee breaks. The likelihood of not running into someone at least once a tour is limited. But Galen, the mud engineer, beats the odds, mainly because he only appears during emergencies.

Tonight is such an emergency and he is in the mud pit room with Clayton, the derrickman. To have both of them in there together is cruel and unusual punishment to each. For they are well over six feet tall and the ceiling of the mud pit room is a bare six feet high between the joists. But, since both are such easy-going, good-natured men, they have not gotten on each other's nerves as hunched over they maneuver around the crowded space. They have dropped the "Excuse me's" of the earlier hours, however, and now simply gently grab one another's rear end and move it out of the way when they want to go anywhere, which is often because the mud is now being weighed every three minutes, so no sooner has Clayton finished one measurement than he has begun the next.

Galen, on the other hand, is walking constantly from the mud pits to the mud room to a workbench by the Halliburton cement pumps where he has set up his portable laboratory. His job is basic. He has to decide which chemicals in what proportion to add to the mud to achieve and then maintain the proper weight and viscosity for the most efficient and safest drilling ahead. His task tonight is to counter the gas that is thinning out the mud and lowering the weight. And he hasn't found the mix. The weight coming out of the hole has been steadily dropping. The parts-per-million of gas has been increasing. We maintained such a steady flow of barite into the mud that Clyde is complaining that we are using the compressors too much to refill the mud room vat and he hasn't enough pressure on the drill floor for drilling. The mud room floor is piled high with empty sacks of caustic soda and Q-Broxin. Pop is white with the barite, John is black with the Q-Broxin. They are both sweating profusely from hauling the sacks. So is Freeman, who, a little worried about being caught sleeping again, has decided to do what he has been told.

But like Clayton, Galen is not excitable. What happens, happens. His only sign of frustration is an infrequent sigh when he hears Hippy report from the shale shaker room. Like Clayton, Galen is a practicing Christian—in his case, born again—and as he puts it: "If anything happens tonight, it will be the Lord's Will. He knows. All I can do is what He tells me," along with what he remembers from his six months in "mud school."

Mud going into the hole . . . fourteen-point-five.
Mud coming out of the hole . . . fourteen-point-zero.

Grady has gone to bed. The box he was sitting on is neatly put away in the far corner of the welders' room. On top of it lies a clean towel and under the towel is the rim. Half the rim has a sheen, the other half is still pocked with tiny dark holes. It does not look like a table yet. The legs are going to be the problem.

Mud going into the hole . . . fourteen-point-five.
Mud coming out of the hold . . . fourteen-point-two.

"I don't know what it is about me. I'm telling you the truth, I can't figure it out myself, but when I fuck 'em they want more. They keep coming back and coming back. Man, I can't get rid of 'em.

"Now, I'm going to tell you, son, when I get back to Baton Rouge, I'll change into my new suit there that I wore out here, I'll get my hair all washed, put on some of the good smelling stuff, get in that Comero, and get me on down to this little place for some dancing. And you know what, friend . . . hold on there a minute . . .

Mud going into the hole . . . fourteen-point-five.
Mud coming out of the hole . . . fourteen-point-three.

". . . yas, sir, I'll tell you there's goin' be half a dozen girls wantin' to just ride around with ol' Hippy here, they goin' want to be just touchin' this ol' hair a mine an' I'm goin' to have to tell 'em 'No, ma'm, you got the wrong dude. I just here for dancin' tonight.' An' they ain't goin' listen because you know what, I probably fucked 'em all one time and they can't forget it. They just want more of ol' Hippy. Ooh, yah." And he does a little back, forward, step to the side and cock the shoulders, a big smile beaming, his perfect teeth glittering through his thin beard, his enormous Afro bobbing, his hands open wide, palm down, signalling that everything is cool.

"Here, man, have a cigarette. The pusher's gone to bed. The rest of the night is ours."

Mud going into the hole . . . fourteen-point-five.
Mud coming out of the hole . . . fourteen-point-five.

"You know what I'm going to do? I'm going to work for these people maybe another six months. I'm going to have a lot of money then and I'm going to tell 'em 'Good-by,' and I'm going down to Jamaica, buy a boat, and I'm going to run dope back here. And I'm going to make so much money I'll quit in a year, buy a place down there—a little place with a little land I can grow stuff on—and I'm going to take it easy for the rest of my days. . . ."

And the little ANALYST reaches over to the table and picks up his glasses by the small coffee pot. Next to the glasses is a book, open flat, its worn spine wrinkled and cracked.

"That's what I'm going to do. You better believe it. I paid my dues. I got shot up and busted up in Nam. I got buddies killed over there and I said then, 'That's it for me. I'm gonna get a job that pays me lots of money and I'm going to take that money and I'm going to get the hell out and take it easy for the rest of my days,' and, pardner, you better believe that's what I'm going to do."

And he picks up the book and with a quick glance at the ticking computer, starts to read.

Mud going into the hole . . . fourteen-point-five.
Mud coming out of the hole . . . fourteen-point-five.

The depth guage by the driller's console reads: "nine thousand, two hundred and fifty-six feet." Despite being shut down for a while Clyde and his crew have drilled nearly five hundred feet. But that is not what they are talking about on the drill floor. Out of the rumor mill has come "the word" that the Schlum-

berger crew is due out on the morning helicopter. This news
lends credence to another rumor that drilling will stop at ninety-
five hundred feet. If that is true, then this could be the last night
of drilling. Stokes' crew will take care of the remaining three
hundred feet. It also means that the rig will be moving. The
rumor mill has fairly well determined that the next stop is dry
dock in Port Arthur and the heavy betting is Brazil after that.

"How good's your Spanish, Catfish?" Clyde yells.

"Senorita Fuckee?" Catfish answers.

Mud going into the hole . . . fourteen-point-five.
Mud coming out of the hole . . . fourteen-point-five.

". . . the Lord will determine it, but I wouldn't mind if it
was a boy," says Galen, the mud engineer.

"What does your wife want?" John asks, hawking a Bear-
like wad of tobacco juice through the grating into the mud.

"You keep doing that, we'll have to give a nicotine weight,"
Galen says. "My wife? She'll be happy with whatever comes."

"When's it due?" Pop wants to know.

"Anytime now . . . she's late already . . . I'd kinda like it to
hold off till I get home if it's taking this long."

"I don' want have no babys, not for a long time," Freeman
offers from his place on the only seat in the mud pit room.

"You ain't ever going to have to worry about that, Freeman," Pop says.

"Why that, ol' Pop?" Freeman asks.

"You go to bed too early," Pop answers.

"Aw, Pop," Freeman grins, "you know I can stay up for
that."

"Stayin' up and keepin' it up, they's two different things,
Freeman. I doubt you can do one, I know you can't do both."

"Shee-it, you dirty ol' man, Pop."

10

Seventh Tour

The drilling has stopped. Skinner's crew started coming out of the hole at 4:30 this afternoon, and Clyde's crew has just finished the job. The hole is at 9,400 feet; 1,000 feet short of two miles; 2,000 feet short of the contracted 11,500 feet. The Schlumberger crew is now beginning to log the well, and for the next few hours we have nothing to do, which is exactly what we all plan to do. As far as this hitch is concerned, we are through. The Schlumberger crew will take the rest of the night to complete its work. Whatever they come up with will have to be discussed with Meta Petroleum in Houston and by the time any decision is made—whether to continue drilling or plug and abandon—we will be off duty, washed up, fed, packed, and ready to leave the rig. In Pop's words, we have had our "last get-up." Even Grady was laughing at supper.

And right on schedule Mother Nature is letting us know that she hasn't forgotten that Wednesday is "crew change." For the past two days the weather has been close to ideal: light winds, at times almost balmy temperatures; seas no more than two to three feet; the sky filled with sun during the day and stars at night. But, according to Dan, the night toolpusher, there's a front due in around midnight, and already the sky is becoming overcast and there is a chill on deck.

It appears that Bullrider has forgotten which is port and which is starboard and where the bow and stern are. This lapse —and there is growing doubt whether it is a lapse or one of Bullrider's many gaps—has caused some concern down in the mud room which we are trying to clean.

There is a natural slope in the mud room floor to the center of the rig, that is, away from the drain hole near the stern wall, and no matter how we sweep and squeegee, the water we have hosed over the equipment persists in flowing back from the hole and spilling over the containing rim and out into the main room where the sacks of chemicals are stacked. We have tried everything to stem it, but it is like herding a bunch of pigs. The water will not go where we shove it and finds a new cranny to slip through and come around behind us so that all we succeed in creating is a frustrating eddy.

But Clayton had an idea which, on the face of it, seemed reasonable but which, in retrospect, has not proved worthy of the effort of thinking it.

"Let's get Bullrider to tip the rig," Clayton said.

"Good thinking, brother," John offered encouragingly.

Bear spat in the water and nodded. Six and a half days on the drill floor have taken their toll. The only reason Bear had been working with us and not sitting on the stairs to the mud pits was that Freeman was with us for a while and a sitting Bear could not criticize a leaning Freeman as "a lazy bastard"—a standing Bear could, especially one with a squeegee in his hands. Now Bear is soaking wet; Freeman has gone with Pop to paint Dan, the night toolpusher's, dog cage; and Bear is looking for a legitimate excuse to sit down again.

Heh-low, ballast control . . . come in, Bullrider.
Yeh, wha' you want?

*Bullrider, you think you can tip this rig backwards so we
can get some water outa the mud room?*

Yeh, I guess so. . . . Ho'd on there.

We are all lined up, poised with our brooms and squeegees,
ready for the water to start its meander aft. He's a smart one,
that Clayton. No wonder he is a derrickman. And the water
begins to pile up against the divide between the two mud rooms
and pour over the containing rim.

"That ain't right," Bear observes.

"You damn right it ain't," Clayton says, running out of the
water to the bitch box.

Ballast control, come in, Bullrider!

Yeh?

*You tipping it forward. Tip it backwards. We got a flood in
here.*

Aw righ'. . . . Ho'd on there.

"What's that silly jerk doing now?" John blurts in amaze-
ment. Now the water is rising to starboard and really is flooding
into the main part of the room. "Doesn't he know his bow from
his stern?"

"You mean his ass from his mouth," Bear shoves by his
chaw.

"Clayton, what you have to tell him is this," John suggests.
"Tell him to stand in front of the panel and say to himself: 'I am
looking forward, my back is the stern, my right is the starboard,
my left is the port.' Then tell him we want his ass lowered."

"He'll never get it," Clayton sighs. "Everytime he turns
around he's going to think he's looking at the bow."

"I wanta know how he knows what end of the bull he's
supposed to be facing," Bear says.

"It probably don't make no difference. For him," Clayton
says and once more picks up the microphone.

Bullrider! . . .

Yeh?

You still goin' the wrong way, Bullrider.
The water ain't goin' out?
It damned well sure ain't.
Well, I be damned. . . . Ho'd on there.
Bullrider, which end of the bull . . . forget it.
Wha's that again?
Nothing . . . lookit, pardner, don't touch nothing. Leave it
like that. It's okay.
Water goin' out now, hey?
No, it ain't, but it ain't comin' all over the mud room nei-
ther, so leave it.
Okay, that's good.

"Start sweepin', men," Clayton says, hanging up the mike. "We got nothing to do anyway."

"That guy shouldn't be allowed out alone," John says, stepping into the water.

Bear spits again, tries to shove himself off the stairs and settles back. "Man, I'm tired. I'm ready for my little bed."

And Pop forgot to close the door to the welders' shop and now there is a fine white dusting of paint down the corridor, neatly carried along the freshly waxed floor by the light draft.

It is too bad because it has taken the bloom off the pride Pop feels for the paint job he and Freeman have completed on Dan's dog cage. And it is a good job. They haven't missed a spot. The cage is spanking white. It is so white (and un-dog-cage-like) that in the gray gloom of the room it appears to have an aureole about it.

Lew is there. So is Dan. The young, red-headed galley hand has walked in. Even Catfish has dropped by. Freeman and Pop are standing to one side, basking in the praise the dog cage is getting, hoping that no one will notice the corridor.

"Uh-oh," says Lew as he starts to leave the welders' room, and steps back, surreptitiously crooking a finger at Pop and Freeman and then at me. "Wait here a minute. I think we got a little clean up job."

We hold our breath as Dan leaves and starts down toward the galley. But the reason Dan is a toolpusher is that he has learned to notice everything, and he doesn't get ten feet along before he looks down, stops, and walks back.

"I think you got a little clean up job on that floor there, men," he says. "Better get it up before it sets."

"I'm real sorry, sir," Pop says quickly, "We didn't know the —"

"Don't matter none, Pop," Dan interrupts. "Just clean it up good."

Pop is now in a flap. He sends Freeman to the paint room for some paint remover, me to the generator room for rags. He races to the laundry room for some detergent and a bucket.

"Ol' Pop's gonna have hisself a heart attack," Lew observes from the doorway of the welders' room.

"I'm sorry," Pop says to the red-headed galley hand. Pop is "sorrying" just about everyone who comes within ear shot.

"That's all right, Pop," the red-head says. "Won't take me long to wax it up again."

"That's a nice kid," Pop says to Lew. "He's the only one of them galley hands that'll fold up the clothes if he has to take 'em out of the drier."

"That's nice to know, Pop," Lew says, grinning. "I know how you need help with that washing of yours."

"My God Jesus, it's high up there."

Dave, the electrician, has just made his first, and he hopes his last, trip to the crown of the derrick. He is sitting in the galley, his second cup of coffee squeezed between his hands, a cigarette dangling from his mouth, his baseball hat pushed way back on his head.

"Hell, that ain't high. You just got to do it a few times is all, then it ain't nothing," says Hippy, who is sitting across from him. Hippy had volunteered to escort Dave to the crown, his reward for which is the right to now needle Dave for being afraid to climb those hundred feet almost straight up a swaying derrick.

"Hell it ain't," Dave answers with a faint smile beginning to form. He is secure in the knowledge that he had to go up, he did, and now he's down, and in the final analysis, that is all Hippy did.

"It's that final twenty feet that got me. They're straight up. You think you're leaning backwards," he continues. "And that wind is beginning to blow. It don't seem that much down here, but up there, it's a Goddamned hurricane."

"You shoulda seen him," Hippy announces to everyone. "He was clutching them rungs so hard I thought he was going to bend them right off. I was about to send down for coffee, he was taking so long."

"What I want to know is why they don't put one of those rims around that ladder. Someday one of the derrickmen is going to miss and fall right off," Dave says.

"You ever seen those derrickmen come down that ladder? Every tenth step they hit. They're flying when they come out of the derrick," Hippy points out.

"We had a derrickman fell out of the derrick on this rig I was working in New Mexico," Clayton, the derrickman, offers. "Decided he didn't want to put on his safety belt, lost his balance hitching up, and down he comes, clutching to the pipe. The ol' driller's about to go into the hole, he looks up and here's this guy coming down. 'What you doin' here?' the driller says. 'Come down for a cup of coffee,' the guy says. 'Well, get the fuck back up there,' the driller says, and the guy goes back up."

"Yeh, well, back in the old days the derrickman used to ride up to the platform on the elevators," Elmo, the company man, chimes in. "We had one guy went up that way, standing up, not holding onto anything, and the driller stops the elevators too fast and down that guy comes only he didn't go back up. Last time they let that happen, I'll guarantee you."

"Now I'll tell you, them derrickmen are crazy. This rig I worked when I was just breaking out roughnecking, we had a derrickman. I don't know how he got there. Couldn't do nothin'

right. This one day we were outa the hole changing bits and he
drops the pipe wrench into the hole. Fifteen thousand feet and
there's the wrench at the bottom. Had to shut the rig down for
two days while we fished it out. Finally we gets it and the driller
takes it, gives it to the derrickman and says, 'Here's your
wrench. You're fired.' Know what the guy does? He goes over
to the hole and drops it back down again. Turns to the driller
and says, 'You wrong. I ain't fired, I quit.' "

"Well, I don't know about any of that," says Dave, the
electrician, "but I'll tell you, pardner, I ain't ever going to be
a derrickman. I been searching all week for why that antenna
up there ain't working and all week I been praying the reason's
down here. But it wasn't and I guarantee you, it fucks up again,
the electrician from the next shift is going to fix it."

"That's it, men. Go on down and get yourselves all pretty.
We'll wake you when the chopper gets here . . . if it gets here,"
Lew promises.

"They fly in this. This ain't nothing to those pilots," says
Pop, the crew change transportation expert.

Right now he is probably correct but the front has hit on
schedule and while there is no rain, the winds have blown up
and the seas are growing white caps. There is no predicting
what will happen when day breaks.

One of the problems with working the night shift is what
to do with the dirty work clothes. A few of the men take them
home, others take a chance and try to wash them. Most throw
them in the locker where they steam and grow mold over the
week and don't lose their odor for days after being washed
again. Pop is a washer and I follow his lead. Freeman is seldom
dirty enough to care. I offer to put them in if Pop will take them
out and dry them. I am coming out of the starboard laundry
room when I am met by four scowling men clearly on the war
path. In the lead is John, the Yankee, followed in order by
Hippy, Jimmy, the electrician, and Jud, the watchstander.

If I were to pick four men from the rig I would not want charging at me, they would do. I flatten myself against the wall as they pass by and stop in front of the second galley hand room.

"Is this it?' Jimmy grunts.

"It's gotta be. Jackson's is the other one," Hippy answers.

"Let's get him," states Jud. "Open it."

John bangs the door open and is rammed through by the charge of the other three. There is a fumbling for the light switch. The light clicks on and off, then stays on and reveals the red-headed galley hand raising his head and drawing his sheet around him.

"What the—' " he mumbles.

"Where's your locker, mother?" snarls Jimmy. "Start looking everywhere," he yells to the others.

What happens is exactly like something out of the movies. Clothes are tossed on the floor, bottles, books, brushes, combs, everything is heaved.

"It's gotta be here," says Hippy. "Keep looking."

"What do you guys want?" the red-head keeps asking. "Just ask me, for Christ's sake."

"Why don't you shut the fuck up," Jud snaps.

"These them?" Jimmy asks, holding up a pair of red underwear.

"That's them," John says, taking them from Jimmy.

"Let's get outa here," Jud says.

"You in trouble, babe," Hippy says. "Your ass is gone." And he shuts the door behind him.

11

To the Bank

Lew bangs into our room at nine o'clock.

"Hey, Pop," he yells, "The chopper's on its way."

"What's that?" Pop groans. His one good ear is flat on the pillow.

"The first chopper's on its way. Just left the bank."

"Are they flying today?" Pop asks. The ear is off the pillow now but apparently is still fuzzy from sleep. It is also a reasonable question under other circumstances. The weather at seven o'clock had been verging on furious. Normally we would not have bothered sleeping, but, concluding that "They'll never fly in this weather" and with the real possibility of having to lay over another day, we had turned in, just in case.

"It's already left, Pop," Lew yells, emphasizing each word.

"What has?" Pop asks.

"The chopper . . . the whirlly bird . . . whoop-whoop-whoop." Lew swings his hand over his head, imitating a helicopter.

"Oh," Pop says, and flicks on the light at the head of his bunk. Lew looks at me, raises his eyes toward the ceiling, and closes the door behind him. I don't have to go on deck to realize the weather that blew in earlier in the morning has quickened. I don't even have to leave my bunk. The room is creaking. The chair by Pop's head slides a short distance across the floor, spilling the ashtray. Pop is impervious. His bunk is now a cradle, but I am wide awake. There's no hope of getting back to sleep.

The day office is filled with milling men. Those who are not peering over someone's shoulder at the flight list are staring out the window at a supply boat tied to one of the anchor buoys. It must have arrived after we'd turned in. The seas pitch it up and then bury it from sight. It looks like the *Juanita Candies*.

"What's the time?" Clyde demands.

"Nine-thirty," Hippy answers.

"It ought to be here," Clyde insists. "When you say it left?"

"They called at seven-thirty," Otis says. "They didn't say if they left, only that one's coming."

"They still oughta be here," Carney, the Barge Engineer, states. "Shouldn't take 'em that long with a tail wind like this. It's going back is going to be bad."

"Maybe they turned around. Maybe they got half way out and chickened and they ain't back yet so the bank don't know if they're coming or not," says Otis, the mechanic.

"What're they flying out—the big one or the little one?" Jud, the watchstander, asks.

"Flying 'em both the first run is what I heard."

"Nah, the little un won't never make it in this weather. They got to fly the big 'un."

"Ain't no matter what they fly. Goin' be some bouncy going back. What's it blowin' now?"

"Gustin' over fifty knots now and comin' up. My money says they'll be lucky to get one flight out here."

"That don't bother me none. Minute that bird hits the deck, I'm gone."

"How'd you get the first flight, Bullrider?"

"Ridin' home with the toolpusher is how."

"Shit."

"Here it comes! See you boys in a week."

"Watch yourself with that door it don't swing back on someone!"

Otis, the mechanic, is on the second flight list. He has made certain. While the scramble was on for the first flight he carefully wrote his name at the top of the second list. Since the first chopper took off he has beat a path between the aft stairwell where he has placed his small suitcase, the galley where he has been chugging small cups of coffee, and the day office where Jake, the crane operator, has taken possession of the flight list.

"When that chopper hits the deck, this little boy's going to be sitting there with his arms open," Otis announces.

"What was that all about last night?" I ask John as we both sip at a bowl of gumbo.

"Shhh!" he says, nodding his head toward the end of the starboard table. The red-headed galley hand is sitting there, his head in his hands, staring down at the table. There is a suitcase at his feet.

"Grady ran him off," John says.

"Ran him off! He was a good kid," and I tell him about the paint on the floor.

"I know, but the guy stole my underwear, brother. That's what we were down there in his room about last night. I got off the tour and went to change and they were gone. I asked him if he'd seen 'em when he was doing our wash and he says, no. But he's the only one that would've.

"I told Hippy about it and Hippy tells Jud and Jimmy overhears and right away says, 'The little bastard's a thief. Let's go get 'em.' I mean, vigilantes. They were ready to lynch him. And sure'n Hell, he's got my underwear in his locker and a couple pairs of pants and a shirt, all washed and folded and stacked in there.

"The kid says he didn't know who they belonged to so he was keeping them safe until he could find their owners. He swears that's the truth. Grady says, 'Guarantee you, he's got some kid brother at home those things are going to fit real fine,' and he goes straight to Jackson and tells Jackson to run the kid off.'"

"Over a Goddamned pair of underwear?" I ask. "Did Grady even talk to the kid about it? What if he's telling the truth? What if he's just stupid?"

"Grady says it doesn't make any difference. The clothes were in his locker. That's all the evidence you need. He says he doesn't want even the suspicion of stealing on the rig. Nip it in the bud, he says. . . . You know what it means? The kid'll never get another job off shore. He's done."

"How'd you like to be sitting out in that thing, Freeman?"

"I damned sure wouldn't. I be sick, I tell ya," Freeman says and from the look on his face it appears he might be at any time.

Freeman and I are standing out by the port stern anchor. The wind is blowing so hard we have to hold onto the railing. The waves are engulfing the supply boat, tossing it around at will.

"Any signs of that chopper yet?" yells Otis, who has worked his way up through the line of men waiting inside the stairwell.

We stare across the heliport past the crane to the northard, searching the horizon for a speck that might be the helicopter.

"Nothing yet," I yell back.

"It ought to be there."

"Ain't," Freeman yells. He turns around and heads back toward the stairs. "I'm gonin' git ol' Pop."

Gulf Star 45. . . . *Come in,* Gulf Star 45.

Yah, this is Gulf Star 45. (That is also Leroy Gaston, Grady's replacement. He came in on the first flight.)

Look, I got a barge here that has broken its tow. Right now its in Block 534 and it looks to me that it's headed down to Block 6—. That's you, ain't it, Gulf Star?

That's a Roger, Captain. You say it's coming this way? Much appreciated, Captain. We'll keep an eye out for it.

Gaston puts the speaker back on the radio in the day office and goes upstairs to the radio room. The radio room has windows on all sides. He checks the chart on the wall, locates Block 534, then looks out the window in that direction.

"Nothin'," he states. "What's more, I don't know what we'd do if there were something. But surer'n hell, we're right in its path."

"Here it comes! Here comes the chopper!"

There's a stampede up the stairwell. The men wouldn't have moved faster if someone had yelled, "Blow out!"

"See you on the bank, brother," John yells pushing past me.

"Fill 'er up with gas," I shout.

"Right on."

"Hey, get that suitcase for me. . . . Geezuz, geezuz! My suitcase."

Poor Otis, the mechanic. There he is at the top of the stairwell and his suitcase is tumbling down. Someone has kicked it over in the upward rush. I stop its fall and run it up to him. Otis grabs it and races away.

As suddenly as the stairwell is empty of men it fills with new faces streaming down the stairs. They jam up at the door, pushing and shoving to get inside.

"Let me in there, man, it's cold out here . . ."

"Hustle it . . . hustle it. I'm goin' be sick. God damn if I ever have a ride like that one again. . . ."

And here comes Otis. He is close to crying.

"They bumped me . . . the sonsuvbitches . . . they bumped me."

Slowly, dejectedly, he climbs down the stairs, suitcase in hand. His eyes are turning red.

"They slammed the door in my face. . . . If I ever find out who took my place . . . the sonsuvbitches . . ."

Hello . . . Gulf Star 45. Juanita Candies. *Can you hear me?*

Yeh, Juanita Candies . . . Gulf Star 45. *We hear you.*

I got a man with a heart attack here. If I come in by the rig, can you lift him off?

I guess we can manage that. Can the man ride the basket?

That's a negative, Gulf Star. *He looks real sick to me.*

I'll send the basket down with a stretcher but I ain't risking my men. You got someone there who'll ride up with him?

Can do, Gulf Star.

Come on alongside, then, Captain. We'll see what we can do.

Tiny Long . . . Tiny, the crane operator. . . . Come up to the radio room, Tiny. Real quick.

Tiny has taken Lew's place. Lew left on the first flight, neglecting, somehow, to make sure his crew was signed up for the second. Wade used to do the same thing. That is why Pop is still sleeping. He is used to being neglected at crew change. I remain always optimistic—and always disappointed. But then Lew is missing all the excitement, except that no one seems to think that a runaway barge and a heart attack rescue are very exciting. If you were to only hear Leroy's tone and not the words, you might think he was asking for a cup of coffee to be brought up.

"Tiny, we got a heart attack on the supply boat," Leroy explains once Tiny arrives. "Think you can lift him off."

"I suppose," Tiny answers perfectly calmly.

"I don't know what condition he's in. Might be bad."

"We'll get him up as long as he don't fall off," Tiny says. "Of course, if he ain't dead when he starts up, he'll be dead when he gets here."

"Do your best, Tiny. What happens, happens."

The supply boat is having trouble. It can't back in. The waves are crashing over the stern deck and busting against the bulkhead of the wheelhouse. And the port crane won't hoist its boom. The mechanic is working feverishly on its engine. The roustabouts are all leaning over the rail watching the supply boat. Tiny is sitting in the crane, his feet up on the window sill. He is smoking a cigarette.

And suddenly there is a roar, the battering of propellors. The last helicopter has snuck in unnoticed to port and is swinging around and landing.

"Let's go," Freeman shouts. Pop and I dash below and grab our bags and charge up the stairwell. By the time we get to the helicopter Otis has already taken his seat and has fastened his seat belt around him. His arms are crossed over his chest in a challenge to anyone to try to remove him.

"That everyone?" The pilot asks over his shoulder.

"They're hauling up a heart attack," Pop says. "He'll probably have to go in."

"How long's that going to take?" the pilot asks.

"They ain't brought him up yet," Pop says, "and the crane's busted."

"Can't wait. They'll have to hold him till tomorrow. Weather's too bad."

And the propellors begin to rev up; the chopper starts to shake and dance on the deck. The tail swings slowly in toward the derrick. The chopper bounces toward the outside rim of the heliport, lifts up, plunges down toward the boiling Gulf seventy feet below, catches some air and sweeps up, banks around the rig over the day office, levels out and heads in toward the bank. None of us look back. Pop lights up a cigarette. Freeman's eyes are closing. . . .

Glossary

abandon When a well is depleted and no longer capable of producing profitably, it is abandoned. A wildcat well also may be abandoned after it has been determined that it will not produce.

barite Barium sulphate, a mineral used to increase the weight of drilling mud.

blowout An uncontrolled flow of gas, oil, and other well fluids from a well to the atmostphere. A well blows out when formation pressure exceeds the pressure being applied to it by the column of drilling fluid.

blowout preventer Equipment installed at the wellhead for the purpose of controlling pressures in the annular space between the casing and drill pipe, or in an open hole during drilling and completion operations. Also referred to as "BOP stack."

boom A movable arm of wood or steel used on some types of cranes or derricks to support the hoisting lines that carry the load.

casing Steel pipe placed in an oil or gas well as drilling progresses. The function of casing is to prevent the wall of the hole from caving during drilling, and to provide a means of extracting the oil if the well is productive.

choke A type of orifice installed in a line for the purpose of restricting the flow and controlling the rate of production.

come out of the hole To pull the drill pipe out of the wellbore.

This withdrawal is necessary to change the bit, or change from bit to core barrel, to run electric logs, to run casing, and for other reasons.

contract depth The depth of the wellbore at which the drilling contract is fulfilled.

Coonass Slang: Cajun; Southern Louisiana natives of French Canadian ancestry

cuttings Fragments of rock dislodged by the bit and brought to the surface in the drilling mud.

derrick Any one of a large number of types of load-bearing structures.

derrickman The crew member whose work station is in the derrick while the drill pipe is being hoisted or lowered into the hole. Other responsibilities frequently include conditioning the drilling fluid and maintaining the slush pumps. He is usually next in line of authority under the driller.

drawworks The hoisting mechanism on a drilling rig. It is essentially a large winch which spools off of takes in the drilling line, and thus raises or lowers the drill string and bit.

driller The employee directly in charge of a drilling rig and crew. Operation of the drilling and hoisting equipment constitutes his main duties.

floorman A member of the drilling crew whose work station is about the derrick floor. On rotary drilling rigs there are normally two floormen on each crew, but three or more men are used on heavy duty rigs.

hopper A form of large funnel through which solid materials may be passed and mixed with a liquid injected through a connection at the bottom of the hopper. It is used for such purposes as mixing cement slurry, that is, mixing clay and oil or water to form a drilling fluid.

joint A length of pipe, usually thirty feet long.

kelly The heavy square or hexagonal steel member which is suspended from the swivel through the rotary table and connected to the drill pipe to turn the drill string. It has a bored passageway that permits fluid to be circulated from the swivel into the drill stem and up the annulus. The kelly transmits torque from the rotary to the drill string.

moonpool The opening below the derrick on a drillship or semisubmersible through which the marine riser and the drill string are joined to the wellhead at the ocean floor.

monkey board A platform on which the derrickman works during the time the crew is making a trip.

mud   The liquid that is circulated through the wellbore during rotary drilling and workover operations.

mud pit   The open tank which is both the originating and terminal point for drilling fluid circulated down the drill pipe and back up the annulus. Pit levl changes are used to detect either loss of fluid to the formation or incursion of gas or salt water into the wellbore. Either indicates a blowout threat.

mud pump   A large reciprocating pump used to circulate the mud on a drilling rig.

rotary table   A circular device with a square hole in the center which transmits torque or spinning action to the kelly or drill pipe, while drilling or breaking out pipe.

roughneck   A driller's helper and general all-around worker on a drilling rig.

roustabout   A laborer who assists the crane operator and performs maintenance and repair tasks. He is a semi-skilled laborer who requires considerable training to fit him for his work.

Schlumberger   Refers to electric well logging. It is derived from the name of the French scientist who first discovered the method. One of the leading companies in this field of operation bears this name. Around drilling rigs throughout the country it is pronounced "slumberjay."

shale shaker   A vibrating sleeve that removes cuttings from the circulating fluid stream in rotary drilling operations.

spudding in   The very beginning of drilling operations of a well.

sling   A wire-rope loop for use in lifting heavy equipment.

stacking a rig   Storing a drilling rig upon completion of a job when the rig is to be withdrawn for a period of time.

string   Refers to the casing, tubing, or drill pipe in its entirety.

toolpusher   A foreman in charge of one or more drilling rigs.

tour   The word designates the shift of a drilling crew and is usually pronounced as if it were spelled t-o-w-e-r. The word does not refer to the derrick or tower.

tripping   Hoisting the drill string out and returning it into the wellbore. This is done for the purpose of changing bits, preparing to take a core, etc.

V-door   An openning in the side of a derrick at the floor level. This openning is opposite the drawworks. It is used as an entry to bring in drill pipe and casing from the pipe rack or catwalk.

viscosity   A measure of a liquid's resistance to flow, such resistance being brought about by the internal friction resulting from the combined effects of cohesion and adhesion. The viscosity of petroleum products is commonly expressed in terms of the time required for a specific volume of the liquid to flow through an orifice of specific size.

washout An excessive wellbore enlargement caused by solvent and erosional action of the drilling fluid.

wellbore The hole made by the drilling bit.

well logging The recording of information about subsurface geological formations. Logging methods include records kept by the driller, mud and cutting analyses, drill stem tests, electric and radioactivity procedures.

wildcat A well in unproven territory. With present day exploration methods and equipment, about one out of every nine wildcat wells drilled proves to be productive but not necessarily profitable.